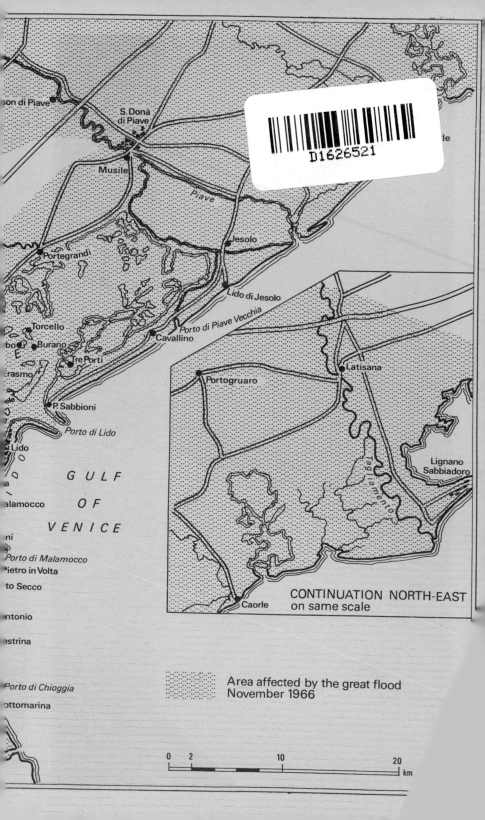

son di Piave

S. Donà
di Piave

Musile

Piave

Jesolo

Portegrandi

Lido di Jesolo

Torcello

Porto di Piave Vecchia

Burano

Cavallino

Tre Porti

Latisana

Portogruaro

erasmo

P. Sabbioni

Porto di Lido

GULF

Lignano
Sabbiadoro

Tagliamento

Lido

alamocco

OF

VENICE

ni

Porto di Malamocco

Pietro in Volta

to Secco

ntonio

estrina

CONTINUATION NORTH-EAST
on same scale

Caorle

Porto di Chioggia

ottomarina

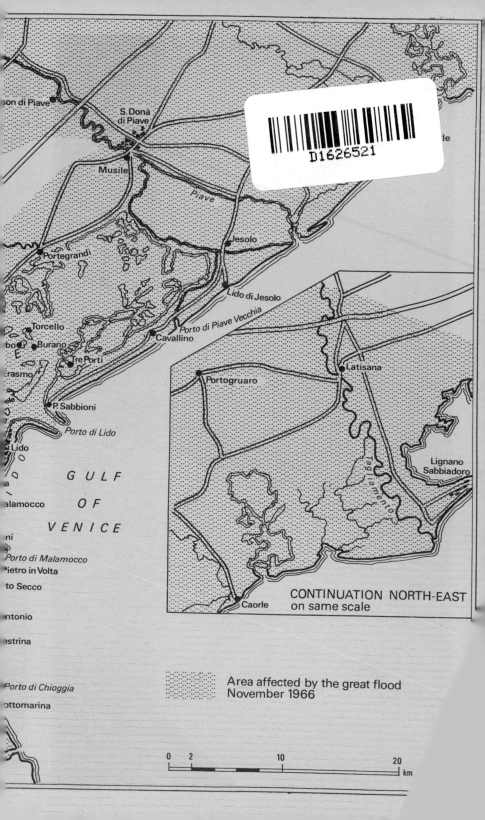

Area affected by the great flood
November 1966

0   2   10   20
|‾‾‾‾‾‾‾‾‾‾‾‾‾‾‾‾‾‾‾‾‾‾‾‾‾‾‾‾‾| km

D1626521

NO MAGIC EDEN

NO MAGIC EDEN

SHIRLEY GUITON

Illustrated by

JOHN LAWRENCE

 HAMISH HAMILTON . LONDON

*First Published in Great Britain
by Hamish Hamilton Ltd 1972
90 Great Russell Street London W.C.1*

SBN 241 02105 7

PRINTED IN GREAT BRITAIN BY
WESTERN PRINTING SERVICES LTD, BRISTOL

TO

MY BROTHER

WHO MADE IT ALL POSSIBLE

CONTENTS

Territory, status,

and love, sing all the birds, are what matter:
    what I dared not hope or fight for
is, in my fifties, mine, a toft-and-croft
    where I needn't, ever, be at home *to*

those I am not at home *with*, not a cradle,
    a magic Eden without clocks,
and not a windowless grave, but a place
    I may go both in and out of.

W. H. AUDEN
About the House

PREFACE

THIS is a light-hearted book and intended to be read as such. But the state of Venice is serious and if it contributes, in however small a way, to some understanding of the problem I shall be very happy.

This book owes so much to so many people. Many friends, or people who became friends, have given me facts and indicated sources and, in that broad exchange of talk which is one of the pleasures of friendship, have prompted, reinforced, modified and sometimes quite disposed of what I thought about the many topics I touch on. I do not know how to thank them all as I ought.

The books to which I owe more immediate thanks are easier to identify. My debt to Giulio Lorenzetti's *Venezia e il suo Estuario* (Libreria dello Stato, 1926) and Horatio F. Brown's *Life on the Lagoons* (Rivington, 1894) appears on every page. I owe particular thanks to Mrs. Mary Lutyens' *Effie in Venice* (John Murray (Publishers) Ltd. 1963) which introduced me to the whole new dimension of Venice during the Austrian occupation.

I would also like to thank the following authors and publishers for permission to use the passages quoted:

W. H. Auden. *About the House*. Faber and Faber Ltd. 1966.

E. M. Forster. *Where Angels Fear to Tread*. Edward Arnold (Publishers) Ltd. 1905.

Gwen Raverat. *Period Piece*. Faber and Faber Ltd. 1952.

James S. Ackerman. *Palladio*. Penguin Books Ltd. 1966.

CHAPTER I

MEETING THE OWNERS

IT was the first day of an international conference. The delegates jostled and pushed about the registration desks; tongues clacked in a dozen pitches and tones; elegant saris disarranged in the crush were fastidiously readjusted and bulky striped Ghanaian robes vigorously re-hitched into place; Moroccans shot the cuffs of their enveloping burnouses with gestures wide as the Sahara and ladies from the Caribbean carried the bright bows of their headgear majestically through the crowd. Latin Americans enfolded their

friends in arms clothed by Savile Row and patted their backs with
a flash of gold rings. An occasional five-gallon Stetson rode above
the crowd on the head of some Texan senator. The Europeans,
sober-suited, a little withdrawn, provided no local colour. They
bore themselves with a modesty befitting those who tend the sacred
flame of civilization and who self-consciously sidestep the question
of where power now veritably resides.

One European was scarcely aware of the bustle, the loud greetings between old friends, the elaborate courtesies between old
rivals, the appraising glance which said: 'What, him again! We
shall have trouble again in the Legal Committee.' For I had a
telegram in my pocket which read:

'Found possible property Torcello come at once.'

The neat round features of an old ally canvassing support for some
ploy dissolved before my eyes, like some slow take in an old-
fashioned film, into the wide stretches of the Venetian lagoon now
bright with sun as Canaletto loved to paint it, now a Guardi of
moody greyish-greens interspersed with the sedgy browns of the
barene. I replied with cagey encouragement: it seemed a good idea,
I would consult my people, I would seek him out and let him
know, we should have another talk. But my mind was busy with
the little sandy island of Torcello set in the back of the lagoon with
a canal dividing the vineyards and the great church rearing up on
the further side.

I do not know how I got away from Paris. I was certainly not
popular with my delegation. But I caught the train to Venice that
night in spite of the ten thousand things to be done.

It was the first of many night journeys between Paris and
Venice. While the house was being rebuilt I had to be in Venice
at least once a month and I have since found that if one is to keep
in touch with a developing vineyard and garden one cannot
manage with anything less. I very quickly discovered that I should
have to travel the cheapest way if I were to make the journey as
often as I needed. Air travel was ruled out not only on grounds of
expense but also because flights are infrequent and are normally

scheduled for daylight hours. Consequently they occupied too much valuable working time both in terms of my office in Paris and of supervision of building and other work in Torcello. Instead I took to travelling by the night train leaving Paris at half past seven on a Friday evening to arrive in Venice at eight o'clock the next morning, returning by the same route on the Sunday evening. A *wagon-lit* costs almost as much as a flight and I soon reverted to the habits of my student days and have since travelled second class *couchette*. With a little ingenuity and some experience one can travel quite comfortably in this way. The trick is partly to command the physical possibilities in an unobtrusive manner. For this the top bunk has advantages. One is out of the way, there is room to sit up and read, one's luggage is conveniently at hand and one controls the lights. A polite suggestion that perhaps the light could now be put out is usually assented to with equal courtesy and prompt action then clinches the matter without causing any shadow of resentment if the switch is at one's elbow. The rule of the game is, of course, that one consents as readily if one has an early bird in the compartment but, in any case, in most trains one now has a little reading light one can use. Again it is wise to arrive a little early to take an opportunity when one's fellow travellers are concentrating on filling up the Customs forms to set the heating controls at a convenient temperature. It has never ceased to amaze me that so few people, while complaining that they are not quite warm enough or far too hot, think to check the controls. Moderation is the golden rule here, too. If one creates a Siberian cold one will be found out and, if one is not, one will undoubtedly perish with cold oneself in the Swiss mountains in the middle of the night; but a slight coolness seems to be regarded by most passengers as a railway vagary for which there can be no remedy.

Basically one's comfort depends on one's fellow passengers. The best travelling companions, without any doubt, are sailors, of whom there are many converging on Venice and Trieste to join their ships. Many years of sharing narrow quarters have taught them a necessary companionable restraint. They stow their belongings tidily in no more, but no less, space than is their due. The common space between the bunks they occupy for the minimum of time but without making themselves martyrs to the comfort of others. Their shoes they put away where they will neither be trodden on nor trip others. They are disposed to comradely chat but bury themselves in a magazine if they see one wants to read. On the other hand they expect one to notice and come out of one's book if there is something they really want to say. They go into the corridor to smoke and they wash themselves thoroughly. They are a lesson in that thoughtfulness for others while maintaining intact their own individuality which is the basis of true gentility. They come in all kinds from all nations and afford one some hope, after the shallow squabbles of international conferences, that man/kind will eventually learn to live together in peace and under/standing.

The next best travelling companions are Italian workers going to and from their homes. When homeward bound, they are weighed down by enormous suitcases bulging with useful pres/ents for cousins even of the remotest remove which are lightly tossed up into the racks when journeying (so sadly) to their work in exile in the misty north. It is no doubt the confined quarters of the poor which has bred in them the habit of not impinging on the reasonable living space of others. Considered from the selfish point of view of one's own comfort their presence means a dis/turbed night because at every frontier on the way they are so often the object of authoritarian coldness, when not worse, about their passports and the provenance and quantity of their belongings and the amount of money they carry. This they bear with the resigned patience of the poor which makes one rage at the deference these same harassing officials pay to one's own bourgeois exterior and manner; minor officialdom acknowledging his paymaster with a humble little bow. Of course this kind of reflection is unfair to the official. The badgered emigrant is seen as a human being. One takes oneself to be a human being. The official in the middle is the outsider in the situation and appears at once as the instrument of oppressive reglementation, as pettifogging in himself and usually as rather cross since frontiers always seem to occur in the middle of the night. He is the one who is not human. But occasionally the façade cracks. One morning at breakfast in the restaurant car a rather aggressive tourist tried to pay his bill in a currency which the conductor was not allowed to accept. He explained the regulations courteously several times but to no purpose. Finally the tourist, a heavily built man of at least six feet four in height, lost his temper, called the conductor names, flung the unacceptable currency on the table and got up to go. But he had to pass the conductor. He stood quite still between the tables, a five foot six Gallic cock balanced to fight. With his head tilted back he looked his huge opponent straight in the eye. After a moment he said coldly and calmly: 'Monsieur, you owe me five francs for your breakfast. I can only accept French, Swiss or Italian money as the train goes through these countries only.' The two glared at one another and for a moment it looked as though there would be a fight. Some breakfasters made furtive preparations to withdraw. Then the tourist looked away, and fished out some French money. The conductor, without undue haste but equally without provocative delay, checked the bill, gave back change, thanked the tourist politely and then stood out of his way. The little man was a minor official again but for a few moments he had made manifest all the dignity and grandeur of France.

The seamen and the workers are seasoned travellers and this too helps to make them better companions than the amateurs. These range from the pathetic to the brash. The former are often timid, tentative and burdened with children and belongings. The largeeyed pale children sit staring at one, bolder than their mothers, not so much crushed, maybe, by time. Overwhelmed by new surroundings and startled by new noises and the landscape charging by, they can be coaxed to a timid smile and even to share a small joke. The old of this type, going to live with a son or daughter mostly as a result of a domestic tragedy, welcome reassurance and seek comfort in precise information: yes, the train does arrive at the Gare de Lyon; yes, it is running on time; surely, the daughterinlaw will be at the station to meet her because her husband will have gone to work by the time the train arrives. Most people travelling on holiday continue to be their ordinary selves, though perhaps a little at sea in a new environment and unhandy at the techniques of travelling comfortably. There is a minority who behave with a selfishness and brashness which one can only suppose is normally kept in check in their own countries and by the society in which they live. These nuisances expect the world to keep to their rhythm and the needs of their brief freedom to govern the conduct of others. They litter the compartment with their belongings. They spread crumbs and orange peel on the floor and talk loudly from bunk to bunk throughout the night.

I do not remember who my fellow-travellers were on that first journey. I was too engrossed in speculations about the house to notice. I knew Torcello well and through the night I reviewed house after house. As we clattered through Dijon I was replanning the interior of the fifteenth-century house that lies on a side canal near the Ponte del Diavolo. As the brakes took hold at Lausanne I was wild with excitement in case it should be the mysterious house by the second bridge opposite Cipriani's restaurant. By Brigue I was gloomily pondering what could be done if it turned out to be one of the tumbledown, four-square barracks that line the canal, that have no privacy and no virtues and are passed each day by thousands of tourists wearily plodding along in the blazing sun from the landing-stage to the great church on the other side of the island. At Milan I decided that the whole expedition was a waste of time. I would have done better to stay in Paris and mind my own business. All my friends, particularly my Venetian friends, had told me that I was attempting the impossible and who could know better than the Venetians? How could my agent have found a house? There were twenty-six houses on the island and I knew them all. Even if, by a miracle, one of them were available it would be bound to cost too much. And then there was no water and no shops on the island. Housekeeping would be impossibly complicated. One had to take a sea journey to buy a lettuce. How ridiculous! And, in any case, what did the telegram mean? 'Found possible property', it said. But the

Soprintendenza ai Monumenti, the all-powerful government department in charge of preserving ancient monuments (and the whole island of Torcello is classified as an ancient monument), would not permit any new building and what would be the use of a 'property' composed of vineyards and fields of artichokes? It was clearly a wild-goose chase. In the early dawn the featureless miles of the plain of Lombardy did nothing to dispel my gloom. At Verona I caught the coffee trolley trundling along the platform and the hot, strong, sweet coffee combined with glimpses of well-known towers induced a less depressed frame of mind. At Vicenza I was altogether too busy trying to verify that you really could not see Palladio's famous Rotonda villa before the train whisked into the railway tunnel to dwell upon my own woes, and by the time I reached Padua a general contentment at being once again in Italy had taken over and I dwelt in a state of euphoria which even the desolation of Mestre could not dispel.

It was drizzling when we arrived in Venice. The specks of quartz in the stone steps leading down from the station glistened in the wet and the broad quay along the Grand Canal was pitted with puddles. I put my suitcase in the left luggage office and took the *circolare vaporetto* line round to the Fondamente Nuove. In the bar by the landing stage I found my agent keeping the wet out with a mildly alcoholic drink. It was he who had sent the telegram and who, from that time on, was to be a true and knowledgeable friend continually going beyond his professional duties to advise and help me. During the thirty-five minute journey to the island of Torcello he told me about the property but I was far too excited to take anything in.

A steady downpour set in as we landed in Torcello blotting out the basilica and the few tall houses. We squelched along the path beside the canal leading from the landing stage to the village *piazza* and just before the Ponte del Diavolo we turned sharp left inland along an overgrown track through the fields. A hundred yards further we turned left again, jumped an overflowing ditch and, bent double, fought our way amid a stench of rotting plants through a jungle of split fruit trees and tumbled down vines, with dripping branches slapping back into our faces. At one point we were brought up short by a yawning well full of brackish water which was dug in the middle of the only path. We slithered round the edge of this death trap, which we later learned was the sole source of water for the vineyard, and found a squat creature with a week's beard and a dripping conical hat peering at us, troll-like, from the bushes.

This was Federico, one of the owners. He led us across a clearing planted with potatoes to a tumbled down cottage whose western wall bulged over the lagoon. The bricks were gnawed hollow by the winds, the roof was half-stripped of tiles and the remaining floor boards sagged under piles of evil-smelling debris. A great fig-tree sprouted where the kitchen had been and the whole was most carefully locked up with a shining new padlock. This was all that remained of a great Cistercian monastery with its chapter house, dormitories and refectories. Among them had stood the church of St. Thomas of the Burgundians. By gradual degrees it had all come down to this. I stood there in the rain sopped through and entranced. My agent gave me the first of many of those anxious looks which indicated that he thought I was on the brink of some folly and launched into a torrent of derogatory comment. I paid not the slightest attention. In Xanadu did Kubla Khan a stately pleasure-dome decree and he had probably started from something like this. It never entered my head that I lacked his resources and that ancestral voices were, if anything, prophesy-ing bankruptcy.

We went back to Venice shivering with cold and slip-slopped along the *calli* to the Hotel Concordia in the Calle Larga S. Marco. We dried off and had a drink and a very late lunch and set out to visit the lawyer who was to look after all the complicated legal work involved in buying land in Italy. He was a courtly old gentleman, sweet-natured and kind and most suitably housed over a flower shop in the Campo S. Salvatore. I very soon realized that beneath this gentle exterior there lurked a highly acute, precise and learned lawyer. He met us at the top of the stairs and ushered us into a room with windows on three sides, two handsome eighteenth-century Venetian canvases of the Madonna and an elegant eighteenth-century table standing on an old Persian rug and surrounded by high-backed carved walnut chairs. Every piece, every ornament was in period. I had been in many rooms whose perfection gave them a cold, unused, museum atmosphere. It was the casual attitude of this Venetian long used to these family belongings which made the room acceptable as a perfectly normal working base. Surrounded by these beautiful things I was introduced to the intricacies of Italian law on the sale of property. The lawyer was more than doubtful about the whole transaction. His doubts ranged all the way from the difficulty of identifying all the heirs and obtaining their consent to the sale to problems relating to the inconvenience of living in Torcello and the suitability of the property. When he was quite satisfied that I was determined to go ahead, he promised all his help and proved as good as his word.

During this discussion I first encountered what was to be a major frustration throughout the negotiations and the building of the house: my lack of knowledge of the Venetian dialect. I had, of course, known that Venetian was a living language but had supposed it to be confined to gondoliers and boatmen and working men and women. Indeed I had equated it with dialects in Great Britain which have lost much of their vigour and individuality under the joint impact of the radio and universal education. In fact the situation is much closer to that of the various kinds of German spoken in the Swiss cantons. I now learnt how generally and with what pride Venetian is spoken in Venice. It provides a focus for local patriotism just as do the various forms of Schwitzerdeutsch. One may wonder whether this strong local patriotism is a happy thing in the world we live in and whether it does not too much encourage among modern Italians that feeling of being separate from, and so, often, superior to, the natives of other cities, which were frequently the rivals in power and wealth and culture of their own town. Italy's great history is constantly treading on the heels of the men of today. Luigi Barzini in his entertaining and rather cruelly perceptive book on the Italians stresses this local loyalty. Since I first read him I have often put his view to the test.

'Where do you come from?' I enquire, pretending that I do not know that I am talking to an Italian, and invariably I get the answer: '*I* am a Roman', '*I* am from Florence', '*I* am a Venetian'. Once only do I recall a man saying simply, 'I am an Italian'. He turned out to be one of the most remarkable Italians I have ever met, with a cool appraisal of his country's possibilities and the chances of European countries working together. But he is not hobbled by the history and civilization (however glorious) of the past. His concern is the future of his country, of Europe and of the world.

The Venetian language embodies reminders of the power, conquests, diplomatic and trading relations of the Serenissima. It took, and gave to the world, the word *arsenal* from the Arabic. It uses straight transfers from the French and treats Italian with an insouciant disregard for the rules of gender which leads to such grammatical frights as *Lio* (*lido*) *Grando* as the name of the new little resort on the lagoon side of the containing *lido* near Cavallino. I have heard gondoliers say 'Vamos!' as they push off a boatload of tourists though the Spanish influence has long gone from Venice and in the lagoon the Spanish 'y' has replaced the 'll' as in Mallorca. Hence, *mia soreya a Torceyo* is *mia sorella a Torcello*. Venice's colonies scattered down the Adriatic coast have left a rich Slavonic vocabulary, and their factories and trading settlements in the Near East many traces particularly in technical and navigational terms.

I now found myself in the vexatious situation of having my affairs discussed by two professional men, acting on my behalf, in my presence, in a language I did not understand. But that was not all. I soon found myself confronted by two other factors.

The first arose from the thoughtful courtesy of all the people I dealt with, beginning with my agent and lawyer, which prompted them to help out my halting Italian by completing sentences for me. The trouble was that these words that were put in my mouth hardly ever expressed what I meant. Often they were diametrically opposed to what I had in mind. I was appalled by the depth of misunderstanding which their kindly meant assistance revealed and by the chasm of incompatible assumptions which yawned between us. What frightened me was that the linguistic barrier proved to be only a surface difficulty and that what was being highlighted was a profound difference in the conception of how life should be lived, of what was important and what was not and the consequent priorities which governed the sort of house one wished to build in which to live the kind of life one wanted to live. This first discussion provided a salutary shock and put me on my guard against the house becoming an Italian house built to suit Italian attitudes to living. It is certainly not an English house (that would be absurd in a Venetian setting) but whether the compromise which finally emerged is a satisfactory one I still do not know. . . .

The second difficulty was that Italy is still a country which thinks that women ought to be protected and have things done for them. Here were two kind professional men (the first of many) who were selflessly devoting their time, their wisdom and their learning to helping me and this, quite naturally, as far as they were concerned, involved taking decisions on my behalf. Quite apart from the fact that I have been an independent, sovereign state for some thirty years now, it is not in my nature to allow people to take decisions concerning me without reference. Yet it seemed boorish to reject a kindness based on a tradition and a form of courtesy which had guided these men all their lives. Later I was dogged by another maddening form of protectiveness. I used sometimes to take a spade and go forth to dig in my own garden (how exhilarating it is to dig in one's own garden) but invariably the dear and invaluable Giulindo, without whom I could not possibly run the vineyard, would appear at my elbow, gently remove the spade from my hand and say expectantly, 'What does the signorina wish me to dig?' What could I do? Stamp and shout and say: 'Go away and let me dig'? Of course not. He was wrapping me round with chivalrous care. So now I have to dig on his day off.

In the larger matter of buying the land and building the house I had to find some way of taking the reins of decision into my own hands and I had to be very diplomatic about it. Heaven knew that I should not succeed at all in this enterprise unless I retained the kindly interest of these men as well as their professional services. Difficulties of language had first revealed the problem and now provided a possible means of solving it. So I began tentatively to edge my way into the conversation, confining myself at first to matters of language, and in the hope of doing no more than delaying the taking of any decision. What did this word mean? What did they mean by that? Could they please explain again exactly the point of this? And of that? With their good nature wedded to the punctilious courtliness of their generation they would stop and do their best to explain.

At first our main language of communication was French. Both my agent and lawyer were of that generation of Italians which learnt French as a first foreign language, but in spite of a few trips to Paris together (recollected in chortling asides interspersed among the technicalities of buying land) this knowledge had become overlaid with time and disuse. I had spoken Italian as a small child (it was, indeed, my first language) and although I could read it fluently my grammar was deficient and my vocabulary elusive, though I readily enough recognized the word I was searching for when someone else suggested it. Later it became our habit to have bilingual conversations: they would talk Italian and I French and we got along much faster. But at the beginning we had not discovered this method and so it was in my halting Italian and their forgotten French that we decided on what tactics to adopt in negotiating with the proprietors of the vineyard. I returned to Paris full of hope.

For many weeks thereafter I struggled with official correspondence with my bank and through it with the Bank of England to obtain permission to transfer abroad the sterling I would need. Luckily my request was made at one of those moments when the economic state of the country permitted some relaxation in the exchange control regulations. I was beginning to think that correspondence about money would never end because the inheritance on which all depended had already been the subject of interminable correspondence and delays with the exchange control authorities in South Africa. It had taken many months to convince them that I had ceased to be a resident in South Africa and that I had quite genuinely lived in England, as my home, for upwards of thirty years.

In the meantime my agent was embroiled with the owners. We had agreed that all the negotiations should be in his hands because the appearance of a foreigner would be bound to double the price. I was, in any case, aware, although my agent was far too polite to say so, that he had no opinion of my bargaining powers. I am sure he is right for all my antique-hunting friends are driven to despair by my impatient disregard for bargaining, and I was very happy to leave things in his hands.

In Italy every part-owner must give his consent to the sale of joint property. In this case we had to deal with seven heirs of the Tagliapietra family. They were Ettore, pensioner; Maria, known as Eugenia, housewife; Antonia, known as Vittorina, housewife; Fioravante, known as Ugo, smallholder; Giuseppe, gardener; Federico, smallholder; Anna-Maria, known as Elsa, teacher. They were the heirs of Gherardo, known as Angelo. The widespread habit in Torcello of being known by a name other than one's baptismal name seems to spring not so much from desire to change a name one may dislike, as from the fact that there are not more than three or four family names on the island. Consequently there are a great many Tagliapietras, all related at some remove and all claiming descent from a count of that name. To avoid confusion between, for instance, two persons called Antonia Tagliapietra, one of them became known by another name, which became a necessary part of her formal identification. Patronymics serve the same purpose. Enrico Tagliapietra, son of one of the heirs from whom I bought the land and who worked for me for a short time, is commonly known on Torcello as Enrico di Fioravante after his father. In exactly the same way the painter Piero della Francesca was called after his mother.

When they inherited the property each of the seven families had reckoned on living at ease on the produce of five acres of land and the work of the other six. When they saw that this dream could not possibly be realized there was loud recrimination. Each put the blame on the sloth, ill-will and incredible lack of fraternal consideration of the others, while they themselves had worked night and day to keep brothers and sisters and their unspeakable progeny fed, clothed and lolling about on the quay. They finally changed their tactics and began to consider what sum of money each required to consent to the sale of the property. Since the consent of all was required by law the possibilities of negotiation and recrimination, of offer and counter-offer were endless. They finally decided, entirely without reference to the value of the property, that they would sell it for a price which would produce one million *lire* for each of the owners, which at the current rate of exchange was about £600. There was, some years later, an even more striking example of this strange kind of reasoning when the seventeen owners of a singularly unattractive barrack of a house with a pokey garden and so situated on the canal that it flooded regularly throughout the winter, refused to sell for less than would provide each of the seventeen of them with a million *lire*. They have still received no offers and the house is now falling down. It seems they would rather have it fall down than bring in less to each of them.

This extraordinarily unrealistic, grasping attitude was accompanied by a kind of innocence, for all their hopes and fears, desires and disappointments, needs and wishes, thwarted expectations and well-founded suspicions about one another's motives were poured out in hours-long sessions with my agent. There he sat, silent, behind his desk, the centre of furious rows and point-by-point repetitions of ancient childhood feuds. He let the brew ferment and bubble and sometimes he gave it a stir himself.

In March 1963 he judged the mixture was ready and summoned them all to his office. As it happened I was in Venice that day and to the consternation of his secretary arrived myself soon after they had all assembled. My presence would have given the whole game away and she quickly hid me away in a little spare office giving on to my agent's room. There I sat listening through the door to yet another repetition of the seven families' claims and counter-claims and the bitter raking over of the muck of old family quarrels. For three hours my agent sat quietly listening to the rise and fall of querulous voices, solo, in duets, trios, quartets and chorus, plaintive, pleading and furious with one another. Then he raised his great foghorn of a voice and quelled them; he refuted their arguments, rehearsed the advantages of selling, made a final offer (which was nowhere near their million *lire* a head) and produced the papers for signature. Total silence followed and I wondered what they could all be doing. They were all signing their names and the names by which they were known. Then they all filed silently out of the office.

A moment later they broke out into a rowdy cacaphonous dispute in the *calle* below. All their suspicions of one another were re-awakened. They had every reason to be suspicious. Did they ever discover that one brother had offered, against payment of half-a-million *lire* for himself, to get the rest to agree to diminish the price by half-a-million *lire*? We were never able to make him understand that there was no advantage to us in paying him half-a-million *lire* unless the price was reduced by more than that amount. He still gives me sidelong looks and thinks me ungenerous and given to driving a hard bargain.

I went into the main office and inspected the document. Ettore, Maria called Eugenia, Antonia called Vittorina, Fioravante called Ugo, Giuseppe, Federico and Maria-Anna called Elsa had jointly sold me full and exclusive rights over a vineyard and orchard together with a small peasant's house of a rustic kind and in an ancient and ruinous condition. I felt as pleased and proud as the Venetian ambassadors must have done when they sailed back from Byzantium having negotiated with the Crusaders the mastery of the Most Serene Republic over a Quarter and Half a Quarter of the ancient world.

CHAPTER II

THE FIRST YEAR IN THE VINEYARD

AT the beginning of 1963 I found myself the owner of some five acres of land and a few weeks later went down to inspect what I had bought. Some of the euphoria of that mad, rain-sodden day in November had worn off and my depression grew as I totted up the evidence of thirty years' neglect. The main part of the property, described as a market garden in the deed of sale, was a tangle of old vines and fruit-trees. This piece runs inland in a rough oblong of about four acres in area with the shorter side

fringing the lagoon on Torcello's western coast. The ruined cottage tottered on the containing bank at this end of the land. Even
the banks were decrepit and riddled with rat runs. The drainage
ditches on the landward sides had, on the other hand, long ago
silted up. The remaining area is divided into two pieces. The
smaller piece, divided from the main parcel by a right of way,
fringes the *bonifica* or common and is at present infested with reeds.
Beyond the canal on its other side lies the second piece which
forms an island of about three-quarters of an acre. It is hopefully
described as an orchard in the deed of sale and there is, indeed,
a thicket of self-grown plum trees at the eastern end. The rest is
covered in rank grass and weeds. The banks have not been reinforced for many years and at every equinox the high tides seep
through. Potentially both the smaller pieces are useful agricultural
land but both need more money and care than I have yet been
able to spare. I put both out of my mind to concentrate on re-
habilitating the four-acre piece and providing a house.

I had bought the property on a verbal understanding with the
Soprintendenza ai Monumenti that I should be allowed to 'rebuild
and slightly enlarge' the cottage. I now started to find out what
this meant and, as a first step, to discover what the Soprintendenza
ai Monumenti was and its reputation in Venice.

The Soprintendenza ai Monumenti is a section of the Ministry
of Fine Arts concerned with the preservation of sites and buildings. The English equivalent is the Ancient Monuments division
of the Ministry of Public Building and Works. The Soprintendenza is a very powerful body which exercises enormous influence
on the fabric of modern Italian society by its attitudes towards the
sites and buildings in its care. These are so numerous as to be
intimately interwoven into the daily life of everybody everywhere.
In a very real sense therefore the evolution of social life is influenced
by the Soprintendenza's sense of balance, by its judgment of when
and what to concede to allow elbow-room for the changes required in a technological age, and when to say 'No' to protect the
past from *laissez-faire* modernists on the rampage. It has a difficult
task at any time and, like all institutions, its task is sometimes

made more difficult by the extremists within its own walls. In times of rapid economic and technological changes such as Italy has been experiencing since the end of the war, the Soprinten- denza's task becomes almost impossibly difficult even without the pressures which powerful interests bring to bear.

These problems are seen in their most acute form in an ancient city like Venice, imperilled as it is by natural phenomena as well as industrial development on the immediate mainland. On the one hand there are the extremists who will not tolerate the slightest deviation from the habits and methods of their ancestors. On the other hand there are those who resent the curbs applied by the Soprintendenza and who are all out for freedom to build and develop businesses as best they can. But when one looks at Mestre one wonders whether uncontrolled development is not too high a price to pay in terms of social cost.

What is so depressing is that Mestre was being spawned over a large area at a time when the reconstruction of war damage and the post-war tendency towards better social conditions had given a new impetus to town planning. Mestre must be seen in the context of the reconstructed centre of Rotterdam; the new residen- tial quarter of Praha in Warsaw and Nowa Huta outside Cracow; West Berlin (and not only the new international quarter in the Hansviertel where Gropius and Alvar Aalto among others con- tributed blocks of flats to a complex pivoting on the new Arts Academy) or the Le Corbusier block of flats near the Olympic Stadium; the new tall blocks of flats in Copenhagen standing in a park-like setting beside a lake, or the great new shopping com- plexes in Stockholm; or our own new towns, with their some- times contested views on how the good life should be lived. Even the pedestrian reconstruction of French towns such as Caen should be taken into account. All these reflect an effort to ensure that more people should enjoy more light and air and other amenities which their particular form of society thinks good.

In contrast Mestre is ungainly and inconvenient. It has none of the dignity of a city and has forgotten the peacefulness of the flat pastures on which it stands. It offers few social amenities to the inhabitants of these rapidly run/up higgledy/piggledy blocks of flats. Its whole life is governed by the stench of lorries trailing wavering blue streamers of petrol fumes like the long paying/off pennants of ships returning home after a lengthy commission in far/off seas.

Yet the Venetians are flocking into Mestre. It offers so much more, they say, than the damp dark houses of the poor quarters from which they come. These do indeed seem to lack the gay bustle which, in some eyes at least, redeems the sadness of the slums of Naples or Marseilles. The existence of Mestre, the fact that it should have been allowed to come into existence, is another indication of that inertia which seems to have beset Venice since the days of its greatness and which has helped the city towards its present decay. One can hardly imagine a city of forward/looking pride (Venice has, God knows, enough of the backward/looking kind) allowing the growth of an incipient slum on its doorstep. What an opportunity Venice missed to show its confidence in its own future by creating, as its other half, a fitting twentieth/century town to serve as an introduction to the incomparable grandeur of the city which ruled a Quarter and Half a Quarter of the earth!

There are, of course, good economic reasons for the existence of Mestre and the neighbouring oil/refinery port of Marghera. There is no room for industrial plants and all the paraphernalia of com/ mercial and technological development in the 123 small islands which constitute Venice. There is scarcely room for the seaport which is now an important factor in Italy's international trade. But economic good sense cannot justify a centre whose confusion frustrates the economic ends it was created to encourage and denies to its inhabitants standards which are becoming commonplace throughout Europe. The purpose of economic advance is ulti/ mately to benefit people. This might be refuted by those who find the unplanned muddle of Mestre no bar to their own immediate gain. They might reflect that a similar attitude in nineteenth/ century England irredeemably ruined most beautiful regions in the North and Midlands and, a hundred years after the event, has bequeathed to us problems which are proving a great drag on the prosperity of those areas and the country as a whole.

There are, of course, distinguished individual buildings in Mestre. There is the splendidly austere new church of S. Lorenzo Giustiniani (the architect is the Venetian, Marino Meo, who also did my house), which is hidden away in the confusion and which deserves to be much better known. There is also the occasional imaginative development of flats and shops. The high standard of this individual work, usually by Venetian architects, rather goes to prove the point that Venice could, if she would, have rivalled the elegance in strictly contemporary social terms of, for instance, the Nordweststadt in Frankfurt. In spite of providing a birthplace for Goethe, Frankfurt hardly sets itself up as the home of imagination or arbiter of the beautiful, yet in the Nordweststadt it has created a dwelling quarter with all the amenities, linked by paths and fringed by roads, where children may play safely and housewives shop conveniently, where people of widely different backgrounds, occupations and social needs may live at ease in a setting which is peaceful and pleasant to the eye. Nordweststadt is a sophisticated version of our concept of a new town, having avoided, or perhaps never having been burdened with, that pervading aura of doing-good which gives a rather adolescent air to our new towns. The brief given to the English planners must have run something like this: It is the English way of life to live in small individual houses in winding roads lined with flowering trees (but not the laburnum which we all know is Victorian) and the people shall therefore be given them. It is good for people to have little front gardens for flowers and little back gardens for veg and they shall therefore have them even if a brief railway journey through the suburbs of any large town provides evidence that most people do not want to bother with a garden. The result is not the fault of the planners. They have done admirable things with their brief and one reason for their success is undoubtedly that most of them agreed with its basic concepts. Consequently, the new towns are most boring places to live in for anyone who does not fall into the pattern of life for which they were conceived. Unfortunately this pattern

excludes most very young people whose rapid turnover of enthu-
siasms cannot be catered for within the limits of such narrow
planning.

The same sturdy *laissez-faire* mentality that has wrecked Mestre
seems to have been at work in the tawdry development of the
seaside resorts which crawl along the narrow beach all the way
from Sottomarina, down at the Chioggia end of the lagoon, to
Jesolo and towards the frontier with Yugoslavia. Here innocent,
bedazzled, package-deal tourists are cribbed in hotels built eyebrow to eyebrow along the beach, presenting their meagrely
balconied bosoms to the sea and their equally exiguous rears to the
next, and lesser, row of hotels which front commodiously, for sea-
food and spaghetti, on the long, thin, noisy thoroughfare which
links this ribbon of resorts. The only escape is into huge caravan
and tented camps interspersed with the resorts all along the coast.

One hopes that tourism, which is an indication of easier and
more leisured conditions, has come to stay and that more people
will be free and able to travel. One might have thought that the
protection of the natural amenities of the sites which attracted
the tourists in the first place was an essential. Tourism is one of
the main industries of Italy and of the Venice region. Along
the Adriatic coast, at any rate, Italy has not done as well as the
Greek tourist industry where government money and control
usually gave a strong lead towards planning and adopting suitable
contemporary architectural styles with a strong local flavour. It is
true that the problem in Greece was easier because tourists are
scattered in smaller groups over many islands. In Mykonos, for
instance, where tourists throng the port and the neighbouring
alleys, the architectural evidence of their presence in the form of
buildings to house them is most discreet and usually admirably
planned even if it does not always match the spare elegance of the
Hotel Xenia in Delphi which manages both to melt into the hillside and assert its own individuality as quite different from, yet
directly descended from, the splendid ruins round the corner.

The Adriatic seaboard is flat and featureless and does not
provide the immediate appeal of an indented rocky coast such as

the Greek. It does have vast stretches of beach sloping gently out to sea and very safe bathing. Children particularly enjoy the freedom these conditions give. Much more could have been done to enhance these attractions. An imaginative start was made. At the beginning of the century the authorities of the little market town of Cavazuccherina (which had acquired its rather ungainly name from the canal cut in 1537 by the engineer Zucchero to reunite the old and new beds of the Piave) decided that their town possessed everything needed to develop into a seaside resort. But who could be persuaded to come to Cavazuccherina? Who could pronounce it? Who could remember it even if they had learnt to pronounce it? Someone suggested changing the name. But how? There was the case of the local village of Musile di Piave which greatly disliked its own name and coveted that of its neighbour S. Donà di Piave. Musile began to call itself S. Donà. Its action was much resented by the true citizens of S. Donà who sallied across the bridge separating the two communities and fought, and were frequently defeated, by the disgruntled men of Musile. After many years the mayors resolved the conflict by agreeing to an exchange of names but as S. Donà was, as a name, regarded as a more valuable property the contract specified that Musile should pay rent for its new name at the rate of one cockerel annually. Every year the bridge over the Piave, which now joins the two communities is decorated and a line painted across its exact middle. On St. Valentine's day the two mayors, accompanied by their village bands, meet on the bridge and the cockerel is handed over the line. A barrel of wine is broached amid speeches, dancing and general rejoicing. I have never discovered what happens to the cockerel. This deal has turned out greatly to the advantage of the new S. Donà for bit by bit the population followed the more prestigious name and settled on the other side of the river. It is the people of the new Musile who generously gave away their name, who attach the most importance to the traditional ceremony. A microcosm of the attitude of the British and French towards their empires? Clearly the authorities of Cavazuccherina who were only seeking the economic advancement of

their township and had no objection, *per se*, to the name by which
it was known, did not wish to get involved in all the difficulties
which had almost proved disastrous to Musile and S. Donà. They
had a better idea. There were many Roman settlements in the
area: Aquileia, Grado, Heraclea and many others. Had there, by
any chance, been a Roman settlement where Cavazuccherina
now stood? Enquiries proved that indeed a settlement had stood
there and that its name was Equilius. This, modified over the
years to suit the Italian ear, had given Jesolo which had been a
flourishing town from about A.D. 700, had risen to be a bishopric
and then dwindled to cornfields by about 1430. The market town
abandoned the name of Cavazuccherina and it is by the name of
Jesolo that it and the resort which grew up within its boundaries
have since been known.

It is a great pity that such an imaginative start should have been
allowed to lose impetus. Now the tourist industry along the
Adriatic, like the Costa Brava and other haphazard tourist
development in Spain, looks like killing the goose which lays the
golden eggs. For should the fashion for lying flat one one's face
on a beach for a fortnight each year give place to another, or
should a more sophisticated generation demand an unspoilt setting for its holidays, people will desert these great holiday conurba-
tions. What horrible slums they will make. Italy, France, Spain
and other countries which have allowed their coasts to be ruined
will be the losers for ever. And England also. Who can condone
the South Coast or the ruin that is Yarmouth? Conservation is
becoming a passion at present, particularly among the educated
young people in America. We must expect young people in
Europe to follow their lead soon. They are the parents of future
tourists and will want to ensure cleaner, saner conditions than we
are likely to have if present trends continue unchecked.

Marghera, the great new oil port, refinery and heavy industry
complex adjoining Mestre, presents problems of a different kind.
If the new interest in conservation is likely to change tourism by
insisting on the preservation of natural amenities, it is conservation's other aspect, which is the growing alarm about pollution,

which is likely to affect Marghera. This expanding system of refineries and port installations has the justification of promoting the economic wealth of Venice and the *terra firma*. This it certainly achieves, but again one must consider what can be done to miti/ gate the cost in social and ecological damage. Marghera has a certain grandeur. From my house at the other end of the lagoon, the flare of the waste gas flickering through the haze lends a point of mystery as well as a Turneresque beauty to the scene. This does not alter the fact that the industrial plants have been partly responsible for creating the haze. Much work remains to be done on the effects of acid and other impurities which pollute the air the Venetians daily breathe, but it seems very likely that they do eat away the stone and metal of the ancient buildings and statuary of Venice and are a factor in the sudden acceleration in the city's decay. It is perfectly possible to filter the smoke of industrial plants to prevent the spewing/out of harmful elements. The clean/air zones of London, Manchester and many other industrial cities which now enjoy sun instead of smog in their busy centres have clearly shown this. It costs money to instal and use filtering plants and one is glad to know that responsible industries (and in Marghera at least one is state/owned) are considering what they should do to meet their duty to the community.

Industry in Mestre and Marghera is seriously affecting the lagoon in three other main ways. First, industrial processes need a great deal of water and a number of plants are at present drawing water from the great natural reservoir which lies below the lagoon. As nothing replaces this water which is pumped away, the bed of the lagoon is gradually sinking. This is another of the new dangers which beset Venice. It would not be true to say that this is the sole reason for Venice sinking. It has been settling into the water since it was first built, as can be seen from the truncated baseless columns which support the Doge's palace. These were not designed like this but successive raisings of the floor of the Piazza have buried the bases and the downward tapering of the column which their shape demands. Secondly, too many plants having used the water pump it back into the lagoon laden with all kinds of impurities. This combined with sluggish tides is changing the ecological balance of flora and fauna in the lagoon with results which are already serious. Lastly, in order to extend the industrial zone the *barene* or low-lying marshes which fringe the lagoon on that side are being filled up. This restricts and distorts the patterns of the movements of water in that section of the lagoon. I shall come back to the complexity of these water movements, which are improperly described as tides, in a later chapter. Here I shall only note that the interference with the natural rhythms of water movement by engineering works such as filling in the *barene* and by excavating a new deep channel to bring very large oil tankers directly from the open sea to the extended oil wharves at Marghera are thought by many Venetians to present new hazards to the life of Venice and the lagoon which must be assessed and, if necessary, eliminated or mitigated.

All these problems are of concern to the Soprintendenza ai Monumenti as they touch the actual preservation of the sites and buildings in its care. The main question remains as to whether the Mestrisation or Jesolofication of the lagoon will in the long run pay off in terms of the development of Venice and the *terra firma* or whether it would not be wiser for the Venetians to consent to some surrender of their personal sovereignty over their individual affairs to promote the long-term prosperity of all. This is a problem not unfamiliar to the six members of the Common Market and the candidates for admission to it who hesitatingly advance, with many panic-stricken leaps back into the cosy kangaroo pouch of their nationalistic past, towards a common approach to problems not so very dissimilar to those of Venice.

The whole of the island of Torcello is scheduled as a protected site by the Soprintendenza. No brick can be laid or change made in the landscape without its permission. As proof of this you can see just by the *vaporetto* landing-stage the brick foundations of what was to have been a little bar. The Soprintendenza disapproved of the shape, size and nature of the building and stopped the project. In the summer it is now decently covered with a tangle of weeds and little self-sown acacia trees and no longer presents a

glaring warning of what happens to those who beard the Soprintendenza. Its policy in Torcello is to keep it as it is—an island of
vineyards and a few peasant families. The Soprintendenza also
has responsibilities for the conservation of the great basilica and of
Santa Fosca, the eleventh-century parish church, the old building
of the Council which shares the little local museum with the
palazzo of the Archives, all sited round the little grassy square.

This miniature community is all that remains of one of the first
settlements in the lagoon. As the Goths swept over the Alps the
Romans pulled back their legions to protect the capital and left the
little towns situated along the northern edge of the lagoon to fend
for themselves. Some took refuge in a group of muddy islets in the
lagoon called the High Bank or Rialto from which Venice
springs. Others settled further away in Torcello. It seems to have
been a fairly extensive island at that time and provided a living for
the new community. It was also secure and was soon the seat of a
bishop, and a great fortified cathedral with heavy stone shutters,
which are still in place in some of the windows, provided extra
security. As peaceful times returned Venice and Torcello, the two
most powerful cities in the lagoon, thought of expansion. The
first requirement was command of the seaward entrances to the
lagoon. Here the Venetians had the advantage as their city lay in
the main channel of the most easterly of the three entrances, now
known as the Lido gate (*Porto del Lido*). Since the other two
entrances lay to the west of Venice, the Venetians were able to
bottle up the people of Torcello in their end of the lagoon, denying
their fishing and trading vessels access to the sea. In a series of
bitter and destructive little wars the Venetians conquered Torcello. From being the equal and rival of Venice it became a kind
of garden suburb, a place where it was cooler and pleasanter when
the *scirocco* blew.

The decline of Torcello seems to have been due largely to the
prevalence of malaria. The island appears to have been healthy
enough in the early days and the arrival of the malarial mosquito
may have been caused by the silting up of the lagoon and consequent changes in the tidal patterns which no longer kept the water

flowing rapidly in the canals. This great scourge is still present in people's minds in Torcello. If anyone is ill they are still said to have *la febbre* regardless of their symptoms. Malaria was only finally wiped out in Torcello at the end of the war by the American army with their D.D.T. spray-guns.

One might argue that the Soprintendenza's standstill policy with regard to development is simply continuing the earlier work of malaria in depopulating the lagoon. It is more likely that the present more prosperous times make the hardships of such traditional lagoon occupations as fishing unacceptable to the lagoon people who will do all they can to put their children into trades which will take them away from the islands. All this is happening. In Torcello twenty years ago the population was 270. Now it is seventy, and composed mostly of the aging. There are now only two children of school age in Torcello and next year there will be only one as the twelve-year-old will have left. These two little girls have a free pass on the *vaporetto* to attend the school in the neighbouring island of Burano. Torcello's school was closed seven years ago when the dwindling numbers no longer justified the replacement of the retiring schoolmistress. The parish church of Santa Fosca is now served by one resident priest whose rank is that of *archiprete* of the basilica. He can call upon the priests in Burano for help at festivals or during his absences. Seventy years ago Santa Fosca was served by several priests.

The Soprintendenza's policy of keeping Torcello as it is suits me. Like Ruskin, I like it as it is. Unlike him I would like to see a revival of old trades and the grafting of new methods on decayed techniques. I see no reason why the peace and quiet of the islands should not be a prosperous peace and quiet. Torcello as it is also suits my pocket. I am one of the very lucky few to have a house on a protected site and a strict enforcement of the Soprintendenza's regulations, particularly with regard to new building, means that my house cannot but increase in value. So, on all scores I started predisposed in the Soprintendenza's favour.

The next step was to decide what to do about the four acres of vineyard rotting away in the undergrowth. I needed to find out

what was there and the use to which it could be put. I was recommended to an up-to-date young surveyor from Mestre called, as it
happened, Torcellan. I soon discovered that any suggestion that
he was in any way connected with the backwater of traditionalism
from which he derived his name filled him with distress.

First he made an accurate survey of the land and, on one of my
weekend visits, he and I went kicking about among the weeds to
find the ancient granite boundary stones which they concealed.
The result of his calculations so surprised him that he took the
measurements all over again. But there could be no doubt about
it. Since the last survey of the site the tides had nibbled away a
strip of several metres wide on one side of my land but had,
luckily for me, washed up the equivalent on the other side. This
made me ponder over the implications for Venice, and for Italy,
of the fact that whereas cities the world over are tending to develop
from East to West, development here was in the opposite direction.

Torcellan then drew up an inventory of plants with the age,
state of health and the expectation of profit from each one, from
the one barren olive-tree a hundred years old (but it is so beautiful
and, as one of the very few existing in the island, it is called on
every year to surrender twigs which are used hereabouts in place
of fronds on Palm Sunday) through a variety of aged fruit trees to
the vines ranging in age from seventy to three years with an estimated crop of eight pounds of grapes each. Opposite is the inventory of trees and plants. It seemed a lot of radishes.

There could be no economic future in such old stock, unpruned
and neglected for so long. The sensible long-term plan was to root
them all out and replant the vineyard on modern lines but this
would cost a great deal and mean waiting a minimum of five
years for a paying crop from the new vineyard. I needed advice
and began to enquire amongst my friends about an expert.

In the meantime something had to be done about the vineyard.
The vines, poor as they were, had to be pruned and sprayed, if
they were to give even a meagre harvest. At this juncture Federico,
one of the seven heirs from whom I had bought the place, asked

Kind	Number	Age in years	Production in Kilos	
			In theory	In fact
Plums	5	10	20	16
	34	15	50	40
	15	20	20	16
	87	25	15	12
	19	30	5	4
Figs	2	5	3	2–4
	3	10	7	5–6
	2	15	16	8–12
	11	30	30	24
	27	50	20	16
	3	dead	—	—
Pears	21	5	—	—
	52	10	40	32
	4	20	70	56
	2	30	100	80
Olives	1	100	—	barren
Blackcurrants	3	—	—	barren
Quinces	28	—	—	barren
Tamarisks	some	—	—	non-productive
Peaches	5	8	50	40
Cherries	1	20	16	8–12
Vines	34	3		
	369	5		
	266	10		
	108	sick	10	4
	1,045	20		
	380	70		
	4	dead		
	2,206			
Artichokes	258	1½	Not yet producing	
Madonna lilies	1,000	—	1,000	1,000
Radishes	—	—	40 sq. metres	40 sq. metres

to be taken on to 'exploit', as he said, the vineyard. I hired him
on a one-year contract. This was not a successful venture. In the
first place Federico still regarded himself as the owner and
instructions as mere suggestions to be brushed aside if he did not
agree with them. It was not that he was wilfully disobedient, but
the habit of ownership was very strong and he was very obstinate.
If all seven owners were as stubborn as Federico what monumental
rows they must have had about where to plant what. He was also
quite genuinely convinced that any instructions given by a foreigner (and more particularly a foreign woman who lived in a
great city) were bound to be nonsense. He had some grounds for
this attitude for the city dweller in Italy seems to be the most urban
of creatures. One of my Venetian friends, who is also one of my
kindest advisers on many things relating to the house, once asked
me, pointing at a young plant which had just been identified for
him as a tomato: 'They grow underground like potatoes, don't
they?'

The main difficulty was that Federico did not understand the
point of the instructions since his mental horizon was very naturally limited to subsistence farming. When he had worked for himself his object had been to keep his family in food and to put
aside his tiny hard-earned margin of profit to provide the ready
money to buy what he could not produce himself such as candles,
paraffin and boots. He was an old man and he drove himself
relentlessly to achieve what amounted, as far as I was concerned,
to a financial loss. His programme was to pick and pack his
produce overnight into baskets or boxes. He got up at four o'clock
the next morning to catch the five o'clock vaporetto to Venice,
carried his baskets across the town from the Fondamente Nuove
which face the cemetery and Murano to the market at Rialto. He
sold his produce through one of the market commission agents,
collected his money (less the sales tax and commission) and walked
across the town again to catch the vaporetto back to Torcello. By
the time he arrived home it was nearly mid-day. He was exhausted
and fit only to potter about when the day was cooler. The cost of
his fares and a modest coffee, which heaven knows I did not

grudge him, left a few pence profit, which he used proudly to hand over to me as proof of his excellent management of the smallholding. To me, his jaunts to market represented a total loss on his day's wages. In spite of my remonstrations he continued obstinately to take quite uneconomically small quantities, such as two baskets of peas, to sell on Rialto. On one occasion he picked fifty Madonna lilies and traipsed off to market. For some reason he failed to sell them. Greatly upset he offered them to every florist and market stall that he could find, but without success. His failure wounded his pride and his self-respect and he came to me to be comforted. Finally I asked him what he had done with the lilies.

'I got tired of carrying them about so I gave them to the Madonna.'

'Whatever do you mean, Federico?'

'I went into a church and gave them to her.'

'?'

'I just told her in a little prayer and left them on the altar steps.'

The attitudes inherent in subsistence farming remain a problem even in the sophisticated agriculture of Northern Italy. The Veneto is neither as fertile nor as technically advanced as the plain of Lombardy. The fact that it does not possess such rich lands in part explains why the people are slow to accept modern methods. For easy production creates wealth which creates a demand for education which creates an open-minded attitude towards new methods and ideas. Subsistence farming is generally recognized as a sign of underdevelopment. In the sense in which Federico knows it it is different from the subsistence farming of the really under-privileged regions of the world. There are available in Italy the skills and equipment to improve the production of the land. There is a system of marketing within reach of the small producers. What is not very different in the two situations is the mental attitudes which bar the Federicos from improving their skills to climb out of the subsistence level to something a little better. As my holding is so small I am registered as a peasant and I know very well the help from the state available to the small producer in price rebates such as that on fuel for tractors. The object of development is to prepare people to adopt new methods to help them to live more fully the kind of life they wish to lead. So it is as varied as life itself and there are universally recurring elements which foster or impede it. Of those found in a situation such as Federico's, of underdevelopment so close to sophistication, some are psychological and some, like land tenure or the general standard of education available to support the desire to progress, are social problems. The old, like Federico, in any case tend to get left behind if only because they have accumulated a backlog of prejudice which no remedial measures can penetrate.

The capacity to develop has very little to do with the inherent intelligence even of people who know they are backward and wish to progress. It is a much more complicated process than merely to discard an old method and follow the instructions for a new technique. Even if the funds are available and they usually are not, at least in sufficient quantity (for undeveloped regions are, by definition, poor), even if the will to apply new methods exists and the people have enough education to appreciate the advantages of using them, their application may require a radical alteration in a precious and basic tradition for which they are not ready. The attitude towards holy cows in India is a case in point. There is now no technical difficulty in rapidly improving the quality of livestock. But there is a whole complex of religious attitudes which have in time given rise to social attitudes which have to be altered before you can begin to improve the livestock. Nor does development progress at the same rate in the different sectors of human activity. There are areas of tremendous resistance where the process of chipping away the accretions of immemorial tradition is heartbreakingly slow. There are others where an objective and unprejudiced consideration of a problem leads directly to a widespread acceptance of the newest techniques. In India, the use of modern contraceptive methods as a means of checking the family and national calamity of over-population are being adopted at an astonishing rate. It would be ironical if India were to solve the problem of improving the human stock before the cattle.

The myth of the great city also affects development. If one is to credit Dick Whittington and the stories of a London paved with gold, it always has. But now news travels faster, though just as inaccurately as before. The city is seen as a modern place where life is up-to-date and therefore good. The shy acquiescence of simple people in this view resists even the bleakness of life in the shanty towns in which too many of them become engulfed. The countryside is slowly permeated with city attitudes and the Federicos of Europe are the more subject to them because they are closer to the great centres of development. He sees life changing all about him and feels a greater need of money because there is a greater variety of goods which undoubtedly make life easier. So his expectations increase but his attitude towards farming remains solidly a subsistence attitude, partly because he has never known anything else but also because his job with me is the only one he has ever had. For as long as he worked for himself subsistence farming was possible. When wages came into it, it was I who learnt the lesson, not Federico.

The pressure to produce more with what you already possess (which is an industrial and not a subsistence concept) also affects the speed at which development is accepted. It is a question of the availability of hands, as we have seen since the war in England in the development of domestic appliances. For as long as there are hands to do the work no one will invent a less onerous way of doing it. In farming as long as there is a grandmother to help out there will always be an old lady placidly knitting mittens on a little folding stool as she supervises her four cows grazing in a clearing, as I recently saw near Zakopane in Poland. For as long as children do not go to school or are taken away early to help on the farm, there will be goose girls to mind the geese, as I also saw there. And, in parenthesis, I also noticed her father ploughing a smallish field with the biggest, heaviest and most unwieldy tractor which I had ever seen. It was a Russian tractor designed to sweep over the vast plains of the Ukraine and points to another problem of development: which is how to provide suitable tools for the conditions in which they will be used at a cost which the country, the co-operative or the individual can afford and of a kind within the capacity of the local mechanics to maintain.

Land tenure systems have a very important influence on deve-lopment. Sometimes it is the laws of inheritance which, while intended to do justice to all the heirs, have the effect of splitting up holdings into smaller and smaller parcels which finally cannot provide anyone with even the means of subsistence. This was the case with the land I bought. Here, a freehold, which in our Western property-owning society whether of large proprietors or peasants is regarded with an almost sacred reverence, had become a useless possession. No doubt some of the animosity between the owners at the time of the sale, stemmed from the loss of status which it entailed. Sometimes it is the terms on which land is rented from a larger landowner. We tend in England to think rather sentimentally about the excellent material conditions of tenant farmers. But this is a nineteenth-century notion. The tenant farmers were the minority who benefited from the enclo-sures, the grouping of uneconomically small holdings and the transition from a medieval form of farming to more scientific methods. These measures created a vast population of dispossessed peasants who naturally resorted to violence in their resentment. One wonders on what scale this strife would have developed if the industrial revolution had not offered a wage in the mines and factories and a place to live in what were to become the slums of the Midlands and North.

The land hunger of the peasants has been a continuing factor in revolutions in the West from the various forms of 'jacquerie' indigenous in Europe up to the French and Russian revolutions. The element which is common to them all seems to be the obtuse-ness of the property-owning classes, whether landed proprietors in the old days, or the bourgeoisie of today with their shares and business interests who, hugging to their bosoms all they possess, refuse to concede anything to others except under the awful threat of violence. When I heard what some of my French friends had to say about the workers and the students, who were sometimes their own children, during the euphemistically named 'Events of May' in 1968, I could not help thinking that we had not moved mentally very far from the attitude of mind of Turgenev's mother who ordered a serf to be beaten to death. It was not so very long ago (she died in 1850) and we have not changed very much.

In a sense the *mezzadria* or share-cropping system which became widespread in a number of countries, including France and Italy, was an advance in that it provided some security and some rules to guide the conduct of both the landowner and the share-cropping *mezzadrino*. Now it is regarded as a form of exploitation which, given the changing circumstances, it is indeed becoming. When large estates were divided up in this way the holdings were of reasonable size and the product fulfilled the expectations of profit of both landowner and *mezzadrino*, which, in any case, were lower than they now are. New agricultural methods demanded large capital investment and large holdings to bring in the much higher profits which then became possible. This encouraged many landowners to take their land back into their own hands. Others, in the course of time, sold off their estates. Gradually the *mezzadrino* found himself working a holding which often constituted nothing larger than the few acres of a tiny home farm and vineyard. Nowadays he can scarcely make a living, his expectations have increased and he accuses the landowner of avarice and exploitation. On the other hand the proprietor is resentful because his share of the profit is so meagre that he has, for the first time, to subsidise his land out of other revenue. From this it is only a short step to accusations of laziness, dishonesty and communism against the *mezzadrino*. The animosity generated on both sides is dangerous to the community. I have a neighbour whose *mezzadria* of some ten acres has been run for the last thirty-two years by a reasonably conscientious *mezzadrino* and his extremely energetic and hard-working wife and sons. Between the time that the two boys left school at the age of twelve and their call-up for military service things went fairly well. The land was in good heart, the vines and fruit-trees were well pruned and each year a large production of annual crops such as peas, beans and *zucchini* brought in enough money to see them through the winter. Then there was a year of glut when they sold very little, and that at a loss, followed by the great flood which put the land out of commission for about two years. The *mezzadrino* had by now used up all his reserves and was living from hand to mouth. For the work of four people it did not seem much return. The owner declined to alter the terms of the *mezzadria*. After thirty-two years of sweating year in, year out over those acres which were not his own, and making a pittance for himself and a profit for the owner so small as to lead to accusations of dishonesty, he left, as he had arrived as a young man, without a penny. His hatred is really frightening and the landowner's attitude to him no less so. He now has a job as a nightwatchman and his two sons have left the land. The elder, with many a backward glance because he really loved it, has become an electrician. The younger, who was never interested in the land, is cheerfully making his way as a waiter. Since the *mezzadrino* left, the owner has had to spend more than it would probably have cost to keep him working contentedly for the rest of his days on contract ploughing and casual help hired by the day to care for the crops. If I am any judge their sale cannot possibly cover the cost of these services and so everybody is the loser. This is an isolated case and a bad one and there must be many where both sides are reasonable and just. But it is symptomatic of a situation which is steadily deteriorating.

In a situation of this kind both trade union and political pressures are brought into play. It seems to be generally expected that as the political left gains power it will bring in a law giving the freehold of *mezzadrie* to the *mezzadrini* who work them. The expectation of dispossession naturally disinclines owners to spend any money on modernization and so the round of misunderstanding, ill-will and animosity continues. The attitude of the agricultural trade unions is understandable. They try to do their best for their members and the short-term best is clearly to obtain for them the whole revenue derived from the land; but in the long run it is absurd as the holdings are already small and will be subdivided by each inheritance. The Communists in this matter are governed by expediency. They want the rural vote and, in spite of their well-known views on collectivization, they work on the expectation that a promise to divide the land amongst the peasants is likely to work, as, indeed, it did in the south of France in the first elections after the war. Ironically there is evidence that those whose main interest is to prevent the extension of Communist power, particularly in the Far East, also advocate dividing the land among the peasantry as the best way to forestall revolution. In the short term this might indeed be effective in deflecting attention from current miseries but in the long run such a policy would be more likely to breed discontent, for even if subsistence farming could provide a secure expectation of a reasonable living, market outlets in most places are extremely difficult to arrange.

Marketing seems likely to be the next big problem. Having for so many years been buffeted by the experts' predictions of world famine in the foreseeable future, we are now told by the Food and Agriculture Organization of the United Nations that we are on the brink of what is already being called the 'green revolution'. Agricultural scientists have recently bred strains of wheat and rice which produce so prolifically that countries which have hitherto had to import rice (such as Burma) now have a surplus to sell on the world market. Elsewhere experimental farms using new methods and new strains of seed are producing crops vastly superior to those of adjacent peasant holdings. In Gezira in Egypt the difference has recently been as much as 80 per cent. But who is going to buy all this produce if everybody grows a surplus? Not all countries can afford to stock and to distribute surpluses at a loss as the U.S.A. has done, and the need to provide food cheaply will not exist if there is universal plenty. Imagine the attitude of mind of a peasant who has only scratched a living all his life confronted by an unsaleable glut. Imagine, on the one hand, the potentialities for exploiting discontent and, on the other, the possibilities of improvements in every aspect of human activity arising from a universal full belly.

In spite of Federico's local marketing problems the grape-harvest that year brought in 800,000 *lire*. Overall this represented a profit of 75,000 *lire* (about £50). It was the last profit I was to see for a number of years.

At the end of his year's contract I asked Federico to go. The land had to earn its living and it could not do so except by a radical change in management methods. Federico did not understand about contracts and thought himself unjustly treated. He hastened off and, for the first time in his life, joined a trade union which, having gone very carefully into the case, had to explain to its new member that he had not been wrongfully dismissed. So he withdrew from that union and joined its rival, with the same unsatisfactory result from his point of view. He was replaced for a brief period by his nephew Enrico di Fioravante. He is small, neat, dark and graceful. He is the local Don Juan and ne'er-do-well and spends the summer fishing and the autumn shooting duck in the back of the lagoon. He told me once that he used to take Ernest Hemingway out shooting. Like most of his stories this seems to be totally untrue. He spent a few months elaborately concealing that he was doing nothing. But by then I had decided to reorganize the vineyard. Enrico was tremendously annoyed when he was sacked. Federico has forgiven me and gives me grapes out of his garden, but Enrico cuts me dead and is the only person on the island who does not wish me good-day.

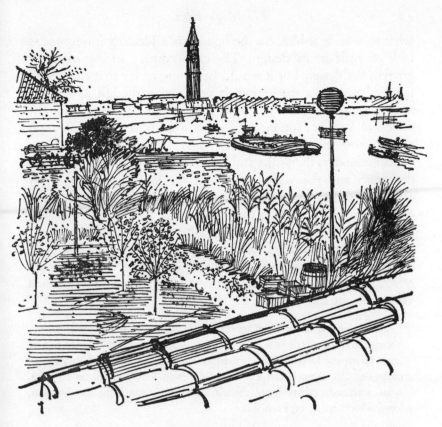

CHAPTER III

NEGOTIATIONS WITH THE SOPRINTENDENZA

THE house and vineyard seemed to occupy both my sleeping and waking hours, for while I deliberately gave my attention to the problem of the vineyard, I seemed to arrive, apparently without conscious effort, at a notion of a house which has not since changed in its essentials.

The essentials were few and, where the house is successful, it is straightforward and candid: the outside makes it perfectly clear what the inside is like. So the house has a pleasing innocence, a lack of guile in its design. The extraordinary setting naturally governs the design but it would be wrong to say that the design takes advantage of the setting. It provides, as though it were the most natural thing in the world, points from which to enjoy the towers of Venice fringing the lagoon to the west, the village roofs and the *campanile* of the great church of Torcello across the vine-yards to the east, the pinks and greens of the Burano waterfront to the south and the wide stretches of the lagoon running away almost, it seems, to the feet of the Dolomites and the high peaks of the Carinthian Alps to the north. It is a plain house. I would have liked it plainer. It is least successful where it has turned out a little too conscious of its antecedents and a little too careful of its architectural manners in fitting into the landscape. *Toute proportion gardée*, it reminds one of Palladio's description of his own Rotonda villa set in the little hills on the outskirts of Vincenza:

> The site is one of the most agreeable and delightful that may be found, because it is on a little hill with gentle approaches, and is surrounded by other charming hills that give the effect of a huge theatre, and they are all cultivated ... And because it enjoys the most lovely views on all sides, some screened, others more distant, and others reaching the horizon, loggias were made on each face.

Internally the house is well-designed. One of its great merits is that it provides the right kind of spaces: space to be social and space to be private, space to be warm and space to be cool, space for books and pictures and the odds and ends with which we bolster our egos and encumber our lives. When time has sorted and com-pressed all these elements and laid a patina of use and habit over all, the house will have become a home.

I did not know, when I began, what vast expense of pain and anxious care would be required to achieve this end. I did not realize either how much of oneself one put into building a house nor how much one learnt about oneself in the process. The work and thought were constant though often below the threshold of consciousness. Solutions to problems would pop up inconsequentially into the routine of one's life—at a dinner party or in the middle of dictating an official letter—and would imperiously require an immediate suspension of the on-going activity while they were checked and mentally filed. Now, whenever I see the eyes of a fellow victim of a cocktail party go suddenly blank and watch him fiddle with one of those limp morsels of food (surely the damp souls of housemaids made manifest), I know that it has suddenly come to him that if he put the stairs in the other way round he would not have to go through the guest room to get to the bath.

Such momentary withdrawals from society are embarrassing but no more. The sort of thing one finds out about oneself entails a much more radical self-appraisal. Take the question of taste. It had just never entered my head to question the certainty of my taste in visual matters. It is true that it had never before been put to the test on anything permanent. I had so far only acquired some books and furniture and a few modern paintings. Perhaps I had been lucky, or possibly had simply remained obtuse about them, but I continued to approve of them because, to me, they seemed perfect of their kind, as the bishop said of the natives of Nepenthe. And if by any mischance I had not, I could very easily have disposed of them without chipping the enamel of my arrogance. It was, I think, the element of permanency, the fact that I would have to live the rest of my life with the results of my mistakes about the house which first upset me. The feeling was brought painfully to the surface by my choice of architect. He was a young man who turned out to have his own rich supply of ideas about whose frightfulness I had no doubt whatever. For some time I lived in a state of most hurtful uncertainty. At one point my *amour propre* sought comfort in pinning all of the horrible plans which emerged on this other fellow. But such subterfuges do not really work. After all I had chosen him and so, by extension, had become responsible for his ideas. It would be very pleasant to think that it was honesty which finally drove me to recognize that some of the bad ideas were my own bad ideas. However that may be, once I had recognized the fact that I, also, had some bad ideas

I had no difficulty in throwing mine overboard and still less in throwing his. In this way I got rid of some ideas which, if carried out, would surely have proved constantly grating reminders of a particularly unpleasing sort of *hubris*. There were many anxieties of a practical order which had to be coped with but none of them were as overwhelming as the idea that I might be led, by my own insufficiency in a field I regarded as of basic importance, to build a house I did not want to live in.

One of the first of the practical difficulties I encountered was distance. While the house was being built I was living in Paris some 700 miles away. This made it difficult to supervise the building operations as closely as a client, for his own satisfaction, should do. At this stage the problem is not one of the quality of the plans themselves. One has scrambled over that obstacle. It is the almost equally important matter of interpreting the plans. I soon learnt that if the details of a house are to be perfect—and a great deal of the final effect and 'liveableness' of a house will depend upon accuracy in detail—the client must turn himself into an architect's assistant and follow closely each step of the building operation, particularly when the major construction work is finished and the plumbing, heating and lighting are being installed. I found this difficult to do from so far away. In practice, I could only come to Venice once a month and then only at the weekend. The difficulty was not so much that something had been done wrong but that subsequent work made it impossible to remedy the original error. For example the plumber installed lavatories of a less good design and quality than those I had myself taken the trouble to go to Padua to choose. They could quite easily have been changed if the mistake had been noticed at once. But by the time of my next visit the bathroom floor and walls had been tiled and any change would have involved stripping them and buying a new set of rather grand and expensive tiles besides paying twice for the time and labour of laying them. In the same way an electric light switch has mysteriously ended up behind the door of the bathroom whose lights it controls. It is true that it is very neatly aligned with the door hinges, which, if anything, makes it worse. The switch could certainly be changed to its proper place but at the cost of stripping the tiles from a whole section of wall. One might argue that super/ vision is the architect's job and so, of course, it is, but it is not reasonable to suppose that a busy man able to spend only a few minutes at a time on the site can match the daily, detailed inspec/ tion of a client's lynx/like eye. It is, after all, the client who is going to live in the house and who will have to endure the nightly exasperation of a light switch out of reach behind a door. It is only he who can correct minor calamities in time. But he must be constantly present to do so and since I could not be there all the time the result is that there exist in my house sources of irritation which I know can never be remedied.

Another source of confusion and strain was my being for a time quite nonplussed by Italian administration. It was not so much that I did not know the regulations. I did not expect to and was prepared to find out. It was that I simply could not conceive why the Italians did things the way they did and it followed that I was constantly surprised, and usually pained, by the kind of obstacles rearing up unexpectedly in front of me. It is extremely unsettling to find oneself blankly uncomprehending about a people one proposes to live amongst. It took me some time to realize that I did not know or understand the Italians and therefore could not understand how they arranged their lives. It also came to me only gradually that, for the first time in my life, I was living in an unfamiliar setting and that part of my difficulty arose from my assuming that I would understand because I always had in England, France and South Africa. Until now I had always been able to draw on that knowledge of the way things are done which children acquire almost by a process of osmosis from family con/ versations and casual references.

I have now gone a little way in knowing the Italians and lack of knowledge at least no longer leads to confusion and delay. Now, at the beginning of any discussion likely to involve public authori/ ties (such as taxation, telephones, water or licences) or traditional attitudes (such as questions relating to money or how you should treat your servants) I ask: what are the rules, what is usually done?

My questions have revealed that a problem exists which is the the reverse of my own, which is the frequent lack of any under/ standing on the part of the local population of a stranger's diffi/ culties. Apart from a small sophisticated section of the professional and wealthy at one end of the nation and the migrants at the other the Italians generally (like the English) have little knowledge of other countries. Consequently my questions, particularly on matters governed by tradition, are often met with a suppressed hilarity and a politely disguised suspicion that I must be a monu/ mental duffer not to know such obvious things. I soon found that I had, in self/defence, to retaliate with accounts of the English way of dealing with the same problem and we all became enchanted at the differences, inconsistencies and illogicalities in our own and one another's national systems.

So I came to recognize that a very good way of getting to know the people you intend to live among is to do business with them. I also learnt a good deal about myself. It requires some experience and an effort towards open/mindedness to observe when it is one's own mental processes which are the delaying factor. I came, eventually, to appreciate the Italians' different method of work and they to appreciate mine. There is, for instance, the difference in timing. The Italians are a hard/working people in spite of their stock image in England. They tend to work intensively in bursts whenever it appears that the work really has to be done. We, on the other hand, tend to work at a steady rhythm for whatever length of time we regard as reasonable and then stop. We give the impression of a rather plodding, joyless approach to our work which is not always true and of a relentless efficiency which one wishes were true more often. The Italian system gives the impres/ sion of an impatient, volatile temperament which is true, but also of disorderliness and indiscipline which is not always true as the work is usually well done and finished on time. Until I realized all this the building of the house drove me into paroxysms of exasperation. I would arrive from Paris and find that nothing had been done since my last visit. Under the spur of my presence the work caught up and was up to schedule by the end of my stay.

The expenditure of nervous energy in these periods of exasperation finally made me ill. When I took myself in hand and calmly surveyed the problem I came to the conclusion that the rate of the work had not really been affected by any pressures exerted by me. My presence, my exasperation had not been a spur. The work had followed its natural, normal, rather staccato rhythm.

The counterpart to this was the willingness and good nature with which all the people I dealt with would turn out to work with me on Sundays. This I found surprising and pleasing and it was of very material assistance to me in planning and building the house. The discussion of all the legal complexities relating to the purchase of the property, of the finances relating to the house and of all stages of both the sketch plans and final working drawings of the house took place on Sunday mornings in the offices of the professional men concerned. They seemed to regard Sunday work as quite normal. There was pressing work to be done so they did it. There would no doubt be a Tuesday or Wednesday with no pressing work to be done so they would not do it. I was delighted and puzzled, given my English attitude towards the division between appropriate times for work and play. I surveyed the professional men of my acquaintance in England and could not see them foregoing their Sunday morning in bed with the papers, or playing with the children or their drink at the clubhouse after a round of golf to turn up at their offices to discuss a client's business. I eventually found an exception, but surely he goes to prove the rule.

All this provided an object lesson in the difficulties of international understanding (that phrase which comes so glibly off the tongue) which was very valuable to me even after twenty years of work in international affairs. It is of particular importance to us to learn to live in harmony and cooperatively with Europe, and also particularly difficult because our geography and our history have led us to develop along rather different lines. Geographically we still feel apart. We are still emotionally islanders and seafarers even though only a minute number amongst us now have anything to do with the sea. The fact that we are a seagirt island still makes us

feel safe and, indeed, in the Second World War the small ditch
which is what the Channel has shrunk to did protect us from
invasion. Would it ever again? Or is the Channel now more of a
natural Maginot Line creating a false sense of security? I only
realized the emotional force of being an islander when I first flew
over the great central European plain on my way to Warsaw. I am
a most unmilitary person but even I suddenly became frightened
as mile after mile of flattish, faintly undulating farm land spread
out with never a check, never a mountain barrier, hardly a river.
Even to my eye it was all totally impossible to defend. All national
boundaries here are artificial lines drawn on a map, matters of
agreement between men, not natural hazards dividing one kind
of people from another kind of people. I began to understand,
for the first time, why Poland had been from the earliest times the
battleground of warring tribes of horsemen, why it had so often
been partitioned, and how it had happened that Hitler's tanks
rolled back the marvellously courageous, under-armed Polish
defence in a matter of days. I thought also of France, naturally
entrenched behind the Alps and the Pyrenees, the Rhine, the
Mediterranean and the Atlantic, with one small vulnerable stretch
between the Ardennes and the North Sea partly protected by the
Maginot Line, and how the sense of security which this gave them
had played them false and was one of the main factors of their
defeat. I thought of ourselves saved then by the Channel and
wondered whether perhaps in the future the feeling of being
separate from Europe which this narrow strip of water gives us
will not play us false and lead us to dwindle away into the off-shore
island of no importance which General de Gaulle foresaw as our
future. We may be able to convince ourselves of the economic
and political necessity of joining Europe but we shall never be of
Europe until emotionally we abandon the concept of being
islanders, this notion which now enables us to live our humdrum,
coy and lazy little lives behind our ditch.

Our history, too, has tended to withdraw us from the cockpits
of Europe in which, in earlier days, we played our rumbustious
part. With the growth of our sea-power and the foundation of

colonies we found we did not need our neighbours in Europe until, in the nineteenth century, proud, rich and controlling the main sources of the world's natural wealth, we became the most arrogant nation on earth. Now we have no empire, no colonies except those poverty-stricken little islands which have no viable political or economic future on their own, and no sea-power. If we are to earn our keep in the new world we shall have to abandon all the emotional baggage which we accumulated in the days of our greatest expansion. This is a very difficult thing to do and it will not be surprising if it takes a generation or even more, particularly if we are to avoid falling into the opposite fallacy which is that we have nothing to contribute to the new world. This is far from true but it is bound to take time to discover which of our qualities we should encourage in these new circumstances and which old attitudes we must abandon for ever.

I found myself mulling over these problems of relations between peoples as I wandered through the *sestiere* of Castello, where the problem hardly exists as few strangers set foot in that end of Venice, or through the first of all the ghettoes in the Cannaregio where the Jews lived with some valuable commercial liberties and a strict personal confinement in towering eight-storey blocks girt by canals, where the problem was tackled by reglementation and segregation: a liberal enough régime for its times and for which the Jews were grateful. I was faced with it every day in negotiating about the house and, more immediately, in finding an architect.

During the next two years I was often to reflect on the strange relationship which grows up between an architect and his client. It has something of the doctor-patient relationship since it also requires confidence and trust. It is a less poignant relationship but more complicated because it deals with subjective matters like taste and hopes and dreams whose management has been less studied than that of the body where trial and experience have led to some knowledge of what is likely to be efficacious. One of the difficulties is that clients are, for the most part, tyros. Very few of us build more than one house in our lifetime and we have no experience to guide us. We do not know what is our proper sphere as clients and what we must leave to the architect.

Looking back on it I now see that my lack of experience showed chiefly in two ways, apart from the major error of choosing the wrong kind of architect in the first place. My two mistakes were having too many set ideas about what I wanted and being too much involved emotionally. The house was to me not at that time a machine to live in (to use Le Corbusier's phrase) so much as the realization of a dream. The first architect I chose was a young man, newly qualified, and I quite deliberately chose him because I thought it would be easier to retain the initiative with a young architect who had not yet acquired too much confidence in himself. I thought, too, that a newly qualified architect would be more likely to have all the new building expertise at his fingertips. My attitude was wrong in every particular. I had overlooked the fact that an architect, like any other professional man, perfects by experience the knowledge he acquired in order to qualify, that he learns what materials and methods are available and suitable in any circumstances by using them, that drawing plans is much assisted by having overcome similar planning difficulties before. If one wants to be brutal about it, an experienced architect gives his client the benefit of the lessons he has learnt at the expense of earlier clients. The only comfort I can derive from the months I spent working with this young man is that someone some day will profit from the experience just as I, later on, probably benefited from who knows what frustration on the part of some anonymous predecessor. A more fundamental and farreaching error was my attempt to retain the initiative. This discounted the fact that any architect worth the name is an artist as well as a practical builder. He cannot work with any satisfaction if he is reduced to the role of a technician, giving advice on loadbearing walls or how many bricks you need to make a terrace. Long after my house was built I read an interview with Jane Drew on, if I remember rightly, the rebuilding at the Elephant and Castle. She said: 'I do not think you can do anything well by doing it on someone else's back. You cannot design a building secondhand.

You can't imagine a consultant poet, can you? What you need is something of quality in the first place.' Looking back on it, I wholly agree. I ought to have thought of it myself before I started but I did not. Nor did I know at the beginning that the client, just like the architect, has his own responsibilities. I learnt eventually that his job is to think out and then to state very clearly the essentials he requires in his house and (if he must) the frills he cannot bear to forgo. The time to argue on details is when the sketch plans have been drawn. A good architect will make them come alive to a novice. But here the client must be very wary. By now the architect will have fallen in love with his own plan and will be talking in enthusiastic terms about one house with the client interpreting him equally enthusiastically in terms of quite another one. In fact you have two conflicting dreams to cope with rather than only one. I am sure that a good client should discuss sketch plans at length with the architect and then brood over them quietly at home to get them fairly into his own perspective. A good architect will encourage him to do just this.

But how does one set about finding a good architect? In the interview I have quoted Jane Drew said that what you need is something of quality in the first place. How do you ensure that you get it? You can, of course, go to Jane Drew or Gropius or Saarinen or whoever is established and well known and has picked a team of able young assistants. But apart from this it seems to me that a large element of luck must always enter into the choice one makes. What happens if one falls upon a perfectly good, competent architect with perfectly sound, justifiable taste which does not happen to be one's own? Does one live miserably in the house for ever? Can one bear to live in it at all? It seems a very chancy business with the likelihood of disappointment all round which is nobody's fault.

In the summer of 1964 I had recommended to me a young architect whose ideas seemed to chime with my own. He had a difficult assignment. He had not only to satisfy me at my most interfering but to satisfy the Soprintendenza ai Monumenti as well and to persuade them to allow me to 'rebuild and slightly enlarge' the derelict cottage in an area scheduled for preservation in its current state. It would clearly require considerable skill in finding sound reasons and presenting them in the form of an acceptable project to obtain their formal approval. And everything hung on their approval.

The architect was a nice young man. He had come up the hard way and had had to combine earning his living with studying for his qualifications. He had not the makings of a genius but had a steadiness and a brisk energy which were appealing. He was neat and exact in his work and had always assembled all the informa/tion I had asked for. He took the commission very seriously. Houses are seldom built in the lagoon and the regulations to protect the site and the monuments of Torcello were so restrictive as to make the building of a virtually new house there unique. I soon came to realize that he regarded the successful completion of this commission as the start of a brilliant career. We started working together on the first sketch plans. I soon noticed with dismay that his tastes ran counter to my own. Some of his suggestions indeed distressed me, perhaps too greatly. I began to recognize that I had chosen him from all the architects in Venice for the wrong reasons and this made me feel very guilty towards him. For the fact that he could not satisfy my taste did not mean that he could not make a good career based on the recom/mendations of many satisfied customers. I felt I could not let him suffer from my failure to think through the situation clearly. This made it difficult for me to express any criticism except in the mildest and most conciliatory terms. Finally, after the exercise of more tact and patience than I knew I possessed, we agreed on a perfectly simple project for an L/shaped house with the short side comprising a pillared *loggia* abutting on the lagoon. I sent him off to negotiate with the Soprintendenza. The plan was rejected on the grounds, so he told me, that a *loggia* of this kind was town architecture not appropriate to the lagoon. I had set my heart on having this *loggia* but when I saw that the Soprintendenza was determined not to have it I repressed the tactless desire to point out that the old customs house, which is now the Bucintoro row/

ing club, on the edge of the lagoon towards Mestre had, in fact, exactly the *loggia* I wished to build. I later came to agree with the Soprintendenza that it would have been neither practical nor appropriate to the surroundings and classified it amongst those bad ideas which I had had such trouble in discarding.

At this juncture a series of meetings kept me away from Venice for nearly three months and I chafed that the negotiations had been suspended by my absence. When I next went to Venice I immediately discovered that my architect had, with the best intentions, submitted not one, but two alternative projects both of which had been rejected. I would not have thought it possible that a perfectly straightforward, simple design could become so overlaid by a mixture of pretentiousness and tween-ness. The plan had acquired, among other things, a narrow cloister that did not even serve to keep the sun out of the rooms. It had little pebble paths dottily winding about and arriving, with oh such fortuitous certainty, at a brand-new stone well-head with wrought iron supports and a bucket expectantly hanging over a drop of fully three feet six inches. The whole, he explained to me, would be orchestrated by flower-beds filled with begonias and other favourites of municipal gardening. It might all be perfectly splendid in some settings but I did not want any part of it in my house.

Much more serious was the fact that he was in a fair way to antagonize the Soprintendenza. Its co-operation was essential. All my efforts would be entirely wasted should they refuse permission to rebuild. Thanks to these plans the project was acquiring the reputation of being a bad project to be rejected at all costs. Though it greatly upset the poor young man to have the negotiations taken out of his hands I decided to go myself to see the Soprintendenza. To get to their offices one goes through the main gate into the Doge's Palace beside the west front of St. Mark's, turns immediately to the right and up to the first floor. One walks along an echoing corridor to some high plain rooms overlooking the Bacino where the Serenissima housed part of its secretive secretariat. Old, handsome and steeped in history, the Doge's Palace is a most appropriate setting for an organization devoted to the conservation of ancient buildings. My architect of course wished to accompany me and it turned out to be a most difficult interview. His lack of experience in negotiating, his desire to please at all costs combined with a nervous apprehension of their displeasure became only too evident when he was faced by the firm, kindly expressed and perfectly reasonable objections of the powerful and experienced men of the Soprintendenza. It also became quite clear that any plan put up by him (and his position had become more difficult as his self-confidence ebbed away) would be bound to be rejected. It was a very bitter moment for him when I withdrew the plans and asked the Soprintendenza to dismiss the project from their minds until I had had time to rethink the whole scheme. It is awkward, in any circumstances, to sack a professional man. In this case I also felt that I was striking away the main prop of a splendid career. But I, too, had a lot at stake and saw no alternative. I had lost a year and some £150 on sketch plans.

CHAPTER IV

THE TRIUMPHANT CITY

THINGS were now at a standstill. I no longer had an architect and no idea how to find one. I had become rather suspicious of personal recommendations both on my own behalf and that of my friends because it seemed unfair to saddle them with any responsibility in such a subjective matter as the choice of an architect seemed to have become. I continued to go regularly to Venice more out of habit than anything else since there was really nothing to do until I had found a new architect.

The time I had spent poring over those dreary sketch plans proved to be useful in an unexpected way. I began to look at Venice and its products with an eye conditioned to some extent by my experience with those drawings. I then saw what was there to be seen from the start had I not been so overwhelmed by the magnificence of the city as to take it at its own valuation. I saw, then, that there was a strong streak of the tawdry in Venice and that the Venetians are capable of juxtaposing, without apparent discomfort, the beautiful, the splendidious, the grandiose, the bad and the vulgar. E. M. Forster summed it up: 'There is something majestic in the bad taste of Italy; it is not the bad taste of a country which knows no better; it has not the nervous vulgarity of England or the blinded vulgarity of Germany. It observes beauty and chooses to pass it by. But it attains to beauty's confidence.'

The bad taste of Venice owes something to the habits of con-spicuous waste of the vastly rich merchant princes, who, by means of lavish display, underscored their power and position. Marco Polo's dinner celebrating his return after twenty years in China is a case in point. He and his uncles came back to find themselves forgotten, their stock at home wasted away and their position dwindled to nothing, though presumably the house had not collapsed into the near slum which can be observed today in the Corte Seconda del Milion behind the church of S. Giovanni Crisostomo near Rialto where he is traditionally supposed to have lived. It seems to have been the classic situation of the forgotten traveller's return but, unlike Ulysses, he did not lay about him with a sword. Instead, he decided to dazzle his contemporaries. The sumptuous dinner cost vast sums. He and his uncles, richly arrayed in rare silks and furs and weighted down with gold and precious jewels, gave them all away to their guests, retired to their private rooms and, at the start of each of the many subsequent courses, returned even more richly clad and adorned to give away, on each occasion, another fortune as mere small mementoes of their long absence from home. That was in 1295. Venice has never forgotten. The Ca' d'Oro on the Grand Canal just above Rialto was in its heyday painted in brilliant polychrome abundantly picked out in gold. It was clearly designed to point the wealth and position of Marino Contarini who built it and it is pleasant to imagine that he got as much satisfaction in creating a work of art as an object of envy, for he must surely have enjoyed his lavish display for itself as well as for the effect it had on his contemporaries. Does it not show how diminished we are by comparison that the enormously wealthy of today hide themselves from our eyes and when they do emerge seek to dazzle us merely with yachts, like the Greek shipowners? It may be that these modest displays of lavish expenditure are considered sufficient to dazzle us, accustomed as we are to the dead level of a drab demo-cracy. Or perhaps a small gesture is regarded as more telling because our taste today runs to plainer lines and less decoration than in the past. If our taste is indeed plainer we should be grateful for the disrepair of many Venetian buildings which masks their earlier flamboyance and makes them more readily accessible to our present attitudes. We might well have some difficulty with the Ca' d'Oro which, in its original state, might have struck us as a thought garish, just as we would have some difficulty with that miracle of beauty the Victory of Samothrace bedizened in many colours as they say she originally was, and brandishing a long golden trumpet like any post-boy on the back of a stage-coach.

Tawdriness attaining to beauty's confidence is best illustrated today in Venice in the objects made of glass which are everywhere displayed for sale. At their best the craftsmen of Murano have always produced beautiful things but how often, through the ages, have they been content with lesser flights. What a depressing procession of technical skill allied to appalling taste is to be found in the Museo Vetrario there. There are still today one or two firms which maintain high standards of traditional and (more interest-ingly) contemporary design but most romp luxuriantly in a welter of vulgarity. They produce ashtrays of contorted shapes and meagre capacity in howling colours. They produce electric lamps with frilly silk shades supported by the gold-bangled arms of little glass negresses insecurely draped in brilliantly striped glass wraps, with coral lips and sparkling pupils set in the flashing white of their eyes.

Artistically, as in so many other ways, Venice takes itself for granted. It is conscious of having reached the heights and is unaware, or unwilling to recognize, that its present achievements in the arts do not match those of the days of political and economic grandeur of the Serenissima. A complacency has set in which is the worst enemy of creativeness. In a sense, Paris is in danger of going the same way. In nineteenth-century painting the impetus came from Paris. Now it is perhaps beginning to live on its reputation as a centre of creativeness. Paris remains a great capital and its atmosphere, which is its own creation, owing nothing to foreigners, still makes it a wonderful place to live in. So it still attracts the young painter into what is now, if not quite a void, at least a territory of only sketchily inspected concepts. For both, complacency about their quite different achievements, living in the past, conservatism are the great dangers. Both, at their best, have displayed an intellectual rigour which they should refurbish and turn upon themselves. They are both at present a little like Queen Victoria, of whom someone said that she was superbly uninterested in defeat. But the Boer War reduced the Queen and the School of Paris is far away and the Serenissima long ago, and being superbly uninterested only in the great moments of the past leads straight to disaster. Venice must learn where she is weak to recapture the energy, the activity, the creativeness and manifold skills, in the arts, in politics and trade which made her great in the past. It is no good exclaiming *Che bello* as the setting sun strikes the façade of S. Giorgio Maggiore, if she allows Mestre to wallow on her doorstep. It is a particularly difficult situation for Venice for she has to find a middle way and somehow to steer between her conservative element who want to preserve her out of existence as a living city and her wild men who are likely to reduce her to an industrial slum. It is basically a matter of mental attitudes, of being prepared to accept new ideas, of pruning away the out-of-date and strengthening what is still valid. Venice is not alone in facing this problem of renewing mental attitudes. It faces all of us from the most advanced industrially to those trying to break into an economically valid rhythm. There is the story of the irrigation system installed in a desert valley at a tremendous expenditure of the United Nations' money and the time of international experts. When all the irrigation dams had been built and the ditches dug and the sluice-gates installed the water was turned into the valley and the farmers thought that the earthly paradise had come. But it had not. Water had been so scarce in that valley that no farmer, once the water had reached his plot, was prepared to shut it off when he had had his share. He let it run on and on creating a swamp in the midst of the arid lands. What was wrong was simply that he had not mentally accepted the notion of plentiful water and of sharing the plenty. We in the West have precisely the same difficulty in changing our mental attitudes. The only difference is that someone should have helped the simple farmers of that valley to grasp the underlying concepts governing an unfamiliar technology. It may be more difficult for him but still difficult for any of us to learn that *tout se tient*, that every change affects everything. 'Only connect', as Forster said. That is a hard thing for any of us to do and the more conservative we are the harder it seems to be.

The majestic nature of Venetian bad taste is encouraged by the keen commercial sense of the manufacturers. They produce what will sell and though many horrible objects are made for the tourist trade they have a ready sale on the Italian market and amongst the Venetians themselves. The catalogue of candelabras with which my electrician sought to tempt me was full of frights designed for them. The only place where I have ever seen one of those wrought-iron spiders with a great lump of glass for a body was sprawling on a Venetian wall. Their commercial sense seems innate and traditional, is one of the characteristics which led them to power, and is found in great things and small. Take, for instance, that long-ago abbess who divided her twenty hens equally amongst her community which, including herself, numbered twenty-one. When each had received her hen the abbess discovered that she had none left for herself so she asked each of her nuns to return to her a quarter of the hen which she had just received.

This commercial sense has been exercised for many hundreds of years in supplying tourists with whatever they might need. Amongst the earliest package tours were the pilgrims bound for the Holy Land and crusading armies off to liberate it. The Vene׳ tians found them all in food and lodging while their relatives rustled up ships to carry them on to the Levant. The trade was large and lucrative and lasted many years. So many forerunners of our modern travelling salesmen came to Venice that the Serene Republic (as nervous of espionage as any Communist country today) hedged them round with restrictions and made them all live together in supervised national communities. The main post office near Rialto housed the German merchants. If you go into the Fondaco dei Tedeschi to buy a stamp today you will be stand׳ ing in the courtyard (now glassed over) which used to be piled high with merchandise, unloaded directly from their ships on to the steps of the Grand Canal, and you can look up at floor above floor of rooms let out at extortionate official rents to those far׳off Northerners who came to buy and sell. The Persian merchants were lodged nearby in the Rio del Fontego in the next canal on the right above Rialto while the Arabs appear to have had a house out at the Madonna del'Orto and lived in amity, as the Prophet said they should, with the Jews who were not far off in the Ghetto.

Other visitors followed in the train of ambassadors who arrived to negotiate treaties, loans and the hiring of fleets from the Serenis׳ sima. Queen Elizabeth was for many years represented by Sir Henry Wotton who seems to have eked out the funds made available by his somewhat cheese׳paring monarch by advising young men in the purchase of paintings. A few visitors, like Montaigne, were disappointed, finding it 'less admirable than he had imagined it'. Most were amazed and enchanted like Philippe de Commines, the ambassador sent by Charles VIII of France, who wrote that the Grand Canal 'est la plus belle rue que je croy qui soit en tout le monde et la mieux maisonée, et va le long de la ville. Les maisons sont fort grandes et haultes, et de bonnes pierres et les anciennes toutes painctes. . . . C'est la plus triomphante cité que j'aye jamais veue'. Then came Thomas Coryiat, inquisitive, wide-eyed, noticing, forever jotting down notes in much the same spirit as James Morris 350 years later. He fills in for us Philippe de Commines' description of the palaces on the Grand Canal which he says are conspicuous particularly for two things, a 'pretty walk or open gallery' which 'serveth for men to stand in without their houses, and behold things', and a pleasant little terrace on the first floor 'that jutteth or butteth out from the main building, the edge whereof is decked with many pretty little turned pillars, either of marble or free stone to lean over'. The capacity to stand and stare and to stroll enjoyably, for which the French invented the word *flanerie*, is one easily caught even today from the Venetians. Coryiat then bustles off to the Piazza which 'ravished his senses' for there you might see 'all manner of fashions of attire', as indeed you still can from the neat black silk trousers of Eastern ladies and fragile saris to the stocky, bundled-up peasants in their local costumes from some remote Yugoslav province who stand about as amazed as ever Coryiat was. In the market he is dazzled by 'the marvellous affluence and exuberancy of all things tending to the sustension of man's life'. We too are amazed at Rialto by 'the wonderful plenty of their meat, fish and fruits . . . as grapes, peares, apples, plums, apricocks: all of which are sold by weight and not by tale: figges most excellent of three or four sorts, as blacke, which are the daintiest, greene and yellowe. Likewise they had another commodity when I was there, which is one of the most delectable dishes for a sommer fruit of all Christendome, namely muske melons'. One might be reading a postcard from one's aunt. John Evelyn, surprisingly, keeps his feet hardly better in the face of this extravaganza of wealth and colour. 'The Merceria,' he opines, 'is one of the most delicious streets in the world for the sweetnesse of it, is all the way on both sides tapistried as it were with cloth of gold, rich damasks and other silks, which the shops expose and hang before their houses from the first floor. They also keep innumerable cages of nightingales that entertain you with their melody from shop to shop, so that shutting your eyes you would imagine yourself in the country, when indeed you are in the middle of the sea.'

Throughout the eighteenth century young gentlemen poured into Venice to perfect their education and, later still, refugees from disapproval at home. All had to be fed and housed and provided with things to buy and enjoy from works of art and mirrors and glass to kickshows and courtesans. No book on Venice would be worth the price without a reference to the famous list of accredited courtesans. So there it is. It seems to have been largely this list and the Venetian love of festivities which gave the city a reputation for libidinous satisfactions and sensual gaiety amongst primmer nations just as it seems to have been the Moulin Rouge and Murger's *Vie de Bohême* which marked Paris down in British eyes as a sink of iniquity in the Naughty Nineties. Both cities seem to have gone rather far, particularly Venice in the eighteenth century, but one is always left with the suspicion that these reputations were carefully fostered as lures for the tourist trade. Paris is still living on it. Every night buses whisk a wellguarded band of tourists from a nude show to a pop singer, from a nip of hooch to a glass of bad champagne and back to the safety of their hotels. Yet there must always have been a good deal of unwicked fun to be had in them both.

The Venetians specialized in routs, masked balls and carnivals and the setting of the Piazza (Napoleon's 'most elegant drawingroom in Europe') entered from the shadowy secret *calli* and left by a swaying gondola receding laden from the Molo, must often have brought on a little madness that tipped the scale too far. John Evelyn thought so. During the Carnival, he says, 'where the women, men and persons of all conditions disguising themselves in antiq dresses with extravagant musiq and a thousand gambols, traversing the streets from house to house, all places being them accessible and free to enter'. 'Abroad', he adds disapprovingly, 'they fling eggs filled with sweete water, but sometimes not oversweete. They also have a barbarous custome of hunting bulls about the streets and piazzas, which is very dangerous, the passages being generally narrow.' So, it was with some surprise that I discovered what a poor innocent little thing the Carnival had become in Venice. It does not have that fierce undercurrent of violence that one senses in the Carnival at Palermo, nor does it galumph with the vulgarity of a down-at-heel Breughel like the Carnival in Basle, nor yet has it that element of private joy which redeems the long rehearsals and organized din of the Carnival in Rio. It is a muted little festival where a few small children got up as Harlequins or Columbines sedately accompany their parents round the Piazza, only very occasionally breaking out into small riots and thumping one another with great coloured balloons, to be dragged off home howling when these burst. I suppose it has not been spruced up for the tourist trade because Lent falls outside the season.

The regattas have retained more of their popular appeal. The modern regattas are all that remain of the ancient sea pageantry of Venice which began in the very early days with the Festival of the Marys which seems to have celebrated a victory over the pirates of the Istrian peninsula out marauding for wives from Venice just as the Romans did amongst the Sabines. It was replaced by the pomp and ritual splendour of official regattas celebrating impor-tant moments in the life of the Republic. The greatest of these was the procession of boats led by the Doge's huge ceremonial barge, the Bucintoro, when on Ascension Day in every year he wed the sea with a golden ring as a sign of Venice's everlasting dominion over the waters. There are still many regattas in Venice and the islands of the lagoon and many of them combine an element of pageantry with an exhibition of rowing skill. The historical regatta which takes place in July is such a mixture. It is composed of large ceremonial barges belonging to each of the six areas or *sestieri* into which Venice is divided, each representing a tableau of historical characters such as that unfortunate Queen Caterina Cornaro who was winkled out of her kingdom of Cyprus by a piece of chicanery on the part of the Venetian Senate and given instead the sovereignty of the city of Asolo in the hills near Treviso where, in fact, she was maintained virtually a prisoner. These barges are accompanied by light racing gondolas painted in gay colours and manned by two standing oarsmen, one on the poop and the other amidships. These craft race against one another on the Grand Canal as well as acting as escorts to the heavy convoy of barges. There are other festivals such as the Feast of the Redeemer celebrated each July to mark the end of a terrible plague in 1576. A great bridge of boats is built linking Venice across the deep channel of the Giudecca and the whole population pours across it to Palladio's church of the Redentore to give thanks for Venice's preservation. In this festival the tourists do not count. It is a purely domestic, spontaneous, inward-looking act of thanksgiving and pleasure.

In 1797 General Bonaparte swept down upon the now decrepit Republic with the cry: 'I will be an Attila to the Venetian state.' The Doge and the Council cowered pusillanimously before the revolutionary fervour of the French armies and the local commit-tee. The Great Council voted away the old constitution of the city by 598 votes against 74. A miserable whimpering end to a thou-sand years of history. Worse was to come. Napoleon ceded the city to the Austrians and a long period of foreign oppression began.

With the change of régime the pattern of tourist response to Venice gradually changed. In the long sad days of the Austrian occupation most Venetians withdrew into their own closed com-munity in resentment against superior forces. A few rebels from their own society, like Byron, were accepted in that part of Vene-tian society which had, from the earliest days, actively resisted the occupation. Byron himself became seriously implicated in an assassination in which the two brothers of his Countess Guiccioli (brave but rash and haphazard plotters) appear to have had a hand. On December 9, 1820, he wrote to Thomas Moore: 'The commandant of the troops is *now* lying *dead* in my house. He was shot at a little past eight o'clock, about two hundred paces from my door. I was putting on my greatcoat to visit Madame la Contessa when I heard the shot. On coming into the hall, I found all my servants on the balcony, exclaiming that a man was mur-dered. I immediately ran down . . . and found him lying on his back, almost, if not quite, dead with five wounds . . . Some soldiers cocked their guns and wanted to hinder me from passing. However, we passed and I found Diego, the adjutant, crying over him like a child—a surgeon, who said nothing of his profession—a priest, sobbing a frightened prayer—and the commandant, all this time, on the hard, cold pavement, without light or assistance, or any thing around him but confusion and dismay.

'As nobody could, or would, do anything but howl and pray, and as no one would stir a finger to move him, for fear of conse-quences, I lost my patience—made my servant and a couple of the mob take up the body—sent off two soldiers to the guard—despatched Diego to the Cardinal with the news, and had the commandant carried upstairs into my own quarter . . . Poor fellow! he was a brave officer, but had made himself much dis-liked by the people. I knew him personally, and had met him often at conversazioni and elsewhere. My house is full of soldiers, dragoons, doctors, priests, and all kinds of persons—though I have now cleared it, and clapt sentinels at the doors . . . The town is in the greatest confusion, as you may suppose.' Byron thought it wise to avoid the resulting police enquiry by removing from Venice ('the greenest island of my imagination' as he called it in another letter to Thomas Moore) to the Countess's family home in Ravenna.

Ruskin, on the other hand, seems to have had few contacts with Venetians except to pick their brains on architecture and comment disobligingly on their characters. His poor wife Effie consoled her frustration as best she might dancing with Austrian officers who seem to have led the usual inbred, self-regardant, easy come, easy go life of occupation garrisons everywhere. The Ruskins were in Venice in 1849 and again in 1852. Yet nothing indicates their consciousness that they were living among a people who had so deeply resented their loss of independence that they had seized the city and expelled the oppressor and fought a desperate months' long siege before the new republic had been defeated. Daniele Manin had successfully led the revolt for political liberty in 1848 and the new republic was not suppressed until August 23, 1849, just a few months before the Ruskins arrived. It is true that Effie in a letter home tells how her English friend Rawdon Brown, who was in Venice during the fighting, used very often to go to his garden (which was not far from the present railway station) 'to view the siege operations and had a kind of seat made for himself in one of the high trees, and when the shots and shells were falling about, he took the best care of himself he could, but it must have been a very dangerous pastime but he seemed to have seen a great deal and got no harm'. One of the siege operations which Rawdon Brown was most anxious to observe was the innovation of flying bombs attached to balloons which were floated over Venice and timed to explode over their targets. They were not a tremendous success and one created fearful havoc in the lagoon island from which they were launched by exploding while it was being made ready for launching. But the interest here is that they were invented and made by one of Effie's most devoted escorts during her stay in Venice, First Lieutenant Charles Paulizza of the Artillery. So Venetian resistance figured in their lives through the rash act of an English friend and the flying bombs of an Austrian officer. There is nothing about the exercise of close Austrian police supervision by censorship, long prison sentences and exile, which bore with increasing severity on the writers, the intellectuals and the liberal professions with whom, one might have thought, John Ruskin would have shown some fellow feeling. But he seems to have noticed as little of the effects of all this on the people of the city whose structure he loved, as he noticed the unhappiness of his wretched young wife whom he expected to lie nightly by his side in maidenly chastity while he yearned over the purities of architecture. One of this selfish man's few comments on the fighting comes in a letter to Effie from Paris on April 28, 1849 when he wrote: 'If they knock down Venice I shall give up all architectural studies: and keep to the Alps: they can't knock down the Matterhorn.' A comment which is startling in more ways than one even if one takes it as a quip to a girl whom he seems to have regarded as an attractive, silly innocent.

Prim disapproval of the people and life of Venice combined with aesthetic appreciation of its setting, fabric and arts is not confined to Ruskin or to nineteenth-century visitors. It dates from much further back. Dürer, who seems to have been a fair-minded man, comments to a correspondent in 1506: 'I wish you were here in Venice, there are so many pleasant people amongst the Italians that it would melt a man's heart . . . On the other hand there are also the most false, lying, thieving rascals that may be found, I think, anywhere in the world . . . They know well their wicked﹍ ness is known, but they do not care . . . But Giovanni Bellini has praised me highly in front of many noblemen.'

This strange mixture of disapproval and admiration may per﹍ haps be due to the fact that most comments on Venice have been made by visitors rather than by residents. Even when the visitor stayed a long time he often remained a visitor, that is to say external to the situation, rather than becoming a resident which involves a sympathetic participation in the life of the place. By a sympathetic participation I do not mean that every aspect of Venetian life should be admired or every fault condoned, but it is noteworthy that most references to Venetians are strictures and this seems to me to be because the foreigner is usually commenting from his own national standpoint. Amongst Englishmen who became resi﹍ dents in attitude are the two remarkable Browns, Rawdon and Horatio. Rawdon Brown, apart from sitting up trees to observe the bombardments in 1848, put together a vast collection of books and documents about Venice which he bequeathed to the British Museum. He put his encyclopaedic knowledge of the Venetian scene entirely at the disposal of John Ruskin when he was writing the *Stones of Venice* and the kindness and friendly guidance stood them (and particularly Effie) in good stead in organizing the comforts of life. He helped with hiring a reliable gondolier for Effie who was characteristically entirely monopolized by Ruskin to ferry him about on sketching expeditions, leaving Effie to walk, which she never seemed to mind, or find a seat in someone else's gondola. He found them a cook and bought their firewood. He is buried in Venice, his body wrapped in the flag of Venice with St. Mark's lion propping up a book. Some thirty years later Horatio Brown, also learned in historical and antiquarian by﹍ ways, began to publish books which provide an affectionate appraisal of the city and lagoon people.

On the whole the nineteenth-century visitor remained resolutely his national self. The Brownings lived a secure English middle-class life on the Grand Canal, Henry James enjoyed with a lingering touch of American self-consciousness his introduction into Venetian drawing-rooms; Proust had Marcel walking on his heels burdened with all the carefully tended anxieties of his Parisian life; Thomas Mann explored the minds of his countrymen in that shocked and sad little story *Death in Venice*; and Frederick Rolfe, brilliant, quarrelsome and pathetic, lived out his rage at English neglect at the pension under the clock in the Piazza when he was in funds or in his *sandalo* moored under a bridge throughout one bleak penniless winter and at all times importuned the fisher boys and gondoliers. They all remain immutably their national selves against the backcloth of the Venetian scene, like the figures in old photographs where one glimpses exotic palms behind the neatly brushed top-hat. They do not want or intend to belong. They are visitors.

Modern conditions underline the tourists' status as visitors. Few of us now have time to take more than a brief holiday though quicker transport makes it possible for us to profit more from these short periods abroad. Until quite recently access to Venice was difficult and took time. Byron left his coach at Fusina and was rowed across the lagoon, a voyage of two or more hours. By the mid-nineteenth century the viaduct carrying the railhead into Venice from Mestre had been built. But the journey still took time and was not worth doing for a brief stay. Even now most visitors spend two nights at least in the city. But although the total number of visitors is still rising (in 1966 a little more than $1\frac{1}{2}$ million tourists came to Venice of whom about $1\frac{1}{4}$ were foreigners and the rest Italian) the pattern is changing. The day-visitor to the historic centre of Venice is a new phenomenon caused partly by better motor access since the building of the autostrada but chiefly by the development of resorts at Sottomarina near Chioggia and Jesolo, Caorle and others along the coast towards Trieste, and the many caravan and camp sites which have sprouted up between them. In 1963 the tourist-nights accumulated by Jesolo alone exceeded

those of Venice proper. This was no doubt because visitors to
Jesolo, with the attractions of a beach and a family holiday, spent
longer there than do visitors to the city. In addition Mestre and the
resorts have room to develop. It is difficult to build new hotels in
Venice itself, and even now there are more beds available on the
adjacent *terra firma* than in the city. All this adds to the wealth of
the region but poses problems for Venice. Venetian tourism is
based on the provision of food and lodging, and family parties who
come for the day from one of the new resort hotels, provided with a
nose-bag, or from one of the camps, provided with sandwiches
put up by themselves, are viewed with some dismay. Venice itself
makes nothing out of these tourists except the profit on an occasional beer or coca-cola and is put to the expense of tidying up
after them. One can feel a certain resentment against those poor
lost souls wandering listlessly along the *calli* and picnicking end-
lessly on the steps of churches and palaces and landing stages and
the arcades of the Piazza. With the development of the surrounding tourist conurbation the problem of the day visitor increases.
What is to be done with him, what facilities should be given him
(there is a woeful lack of lavatory accommodation for such visitors
already) and how are they to be provided within the meagre
profits from what he spends? Perhaps something will have to be
invented for visits to the phenomenon that is Venice analagous to
the small fee we pay to visit such another phenomenon as Stone-
henge or to open-air museums in the Scandinavian countries.
Tourists who spend a night in Venice are already taxed so a day-
tax or entrance fee could not be regarded as an unjust burden.

Another recent change which can be traced back to the increasing number of day visitors is the change in the standard of dressing.
Tourists on a sight-seeing holiday, staying in towns and hotels,
have in their suitcases, as a matter of course, clothes which can
suitably and comfortably be worn in a city like Venice. The
camper and caravanner, and the beach-side holiday-maker, often
have not. The Venetians are rather conventional people and rather
dressy in their provincial way. To them Venice is a city requiring
a certain formality in dress. No Venetian, unless he was on his way

to the Lido, would think of going through the streets in his shirt
sleeves. His wife would not dream of going out except in a spot-
less and newly ironed dress. The Venetians disapprove of the *sans
gêne* with which foreigners treat their city and they regret the increase in the numbers who meander sweatily about the lanes of
Venice in a rumpled dress or drip-dry shirt and none-too-clean
shorts. The Germans, the Austrians, and, unfortunately, the
English seem to be sweatier and more rumpled than most though
perhaps this is simply because there are more of them, a fact confirmed by the figures of these nationalities staying in the adjacent
terra firma. The Americans and the French, on the other hand,
usually, according to these figures, stay in Venice and, as one
might expect, tend more to dissolve into the Venetian crowd,
indistinguishable from the sleek, evenly bronzed Venetian middle-
class which acquires quite a different suntan from the deep,
patchy sunburn of the fisher or boatman burnt black working in
the sun. With a minority of holiday-makers, wearing beach or
camping clothes in the city springs from an almost aggressive
insistence to themselves that they are on holiday. Sometimes it is
simply the discomfort of the unaccustomed heat. I recently saw
one such English couple enjoying the relative coolness of some
steps on the shady side of a church. He had his jacket across his
knees and his waxed guardsman's moustache drooped a shade in
the heat. His rather large wife had one of those blouses made from
two gaily patterned silk squares which, worn outside the skirt, so
obligingly glide from the bosom over the middle bump with
hardly a hesitation. There was an endearing bashfulness in their
manner, a recognition that Venice was a great city, not a fair-
ground. They were doing their best, but would so much have
liked to take off their shoes. One salutes such instances of sensitivity towards the spirit of the city. The day tourist has his problems as well as Venice. He usually has to make an early start and
faces a long day in much higher temperatures than those to which
he is accustomed. In nearly every building or museum he visits as
well as while exploring the town, he has to walk or stand and
there is nowhere for him to sit when he is tired except the cafés

which are probably beyond his means. One wishes that he got
more out of his visit than these circumstances allow and carried
away a glimpse of something splendid rather than a recollection of
sweat and thirst and swollen ankles.

However vexed the Venetians may be with the behaviour of
some of their visitors, they are nevertheless valuable to them in two
ways. In the first place, they bring a great deal of money into the
city every year, about thirty thousand million lire or £20 million.
In the second place the tourists reinforce the Venetians' feeling of
being different. It is an unstated, perhaps unrecognized result of
tourism but one which has its importance. The particularity of
Venetian life grew out of the particularity of their strange and
beautiful city. Their history and the problems they had to solve
gave rise over the years to customs different from other peoples'
and traditions different from those of their next-door neighbours.
So they came to enjoy their differences and they still encourage
themselves to feel different and to maintain old evidences of their
difference. Sometimes they spruce up these evidences with an eye
to the tourist trade but their love of pageantry and ceremony, the
way they cling to symbols and continue ancient institutions, has
much deeper roots. The pigeons are fed every day. It is 'our
custom'. All Venice goes across the bridge of boats to the
Redentore each July. It is 'our custom'. And across another
bridge of boats to the church of Santa Maria della Salute to give
thanks, each November, for another plague which ended in 1630.
It is 'our custom'.

Some are official symbols; some seem to have acquired the
status of symbols by long usage. Take Florian's. To the summer
visitor Florian's may be just a café on the shady side of the Piazza
with a string band thumping out the forgotten tunes of the youth
of all our generations. But go to Florian's in the winter when all
the hundreds of tables and chairs set out in a block have disappeared from the Piazza. The old Florian is resurrected: a symbol
of a conspiratorial past. The atmosphere of earlier days takes over.
This series of secret little drawing-rooms with red plush benches
and swivel-topped tables on stout marble legs, the walls covered

with fading frescoes or tarnished figured mirrors, breathes an air of intrigue and dalliance. Even with modern lighting these rooms are dark and rather spooky. By candlelight they must have been an ideal place for dubious assignations, for the passing of a little note from hand to hand and for the frequent, casual-seeming deliberations of the determined men who plotted rebellion in these rooms under the noses of the Austrian police. Now they are, as they must ever have been, the exchange for knowledgeable, minute and delicious gossip. There are faces that one always sees as one comes in out of the winter mist swirling along the arcade. The old lady, I should rather say gentlewoman, to give the flavour of her confident, ancient breeding with grey hair soberly dressed, who with a hint of a bow recognizes one as another *habitué*, takes a delicate sip of coffee between sentences of a sibiliant conversation (the lilt and the zezeze show she is speaking Venetian) with a younger woman. A daughter? Perhaps. The older lady looks exactly like Veronese's St. Anne (the one that is in Birmingham, as it happens) though the daughter does not come from any picture I can recall. One of the delights of Venice is the persistence of types used as models by the painters. It is a kind of walking catalogue of paintings. There goes a Titian madonna with a glint of copper in her hair walking with that slightly swaying dignity which he would surely recognize, off to the shop at the corner with her market bag. That gondolier came out of Giovanni Bellini and a man I often see eating his dinner in Montin's *locanda* down by S. Trovaso smiles at me from a Veronese fresco every time I go to Maser. The quality of gossip I am sure must be different in Florian's from what it is over the way in Quadri's gilt drawing-rooms or in the little gallery at Lavena's, so convenient for spotting who is taking coffee with whom. It is a question of atmosphere, a marriage between the temperament of the *habitués* and the accumulation of past events. The predominant note, for me, at Florian's is conspiracy, so much so as to stand for it.

If Florian's has become a symbol by usage, the pigeons are now quite official. They are perhaps, along with the gondola, Venice's best-known symbol. Nobody quite knows how they obtained

their status but it certainly began a long time ago. Some say that
it was because the Doge, Dandalo, sent news of his victory at
Constantinople by carrier pigeon. Others that it came from the
custom of releasing a cloud of pigeons from St. Mark's basilica on
Palm Sunday and that this custom itself came from Southern
Russia or Persia. Because of their connection with Palm Sunday
they were protected and have bred ever since in complete im-
munity. I have certainly never known a Venetian eat pigeon.
They are now on the strength and are fed at the expense of the city
promptly at nine and two o'clock each day. This is particularly
welcome to the pigeons during the winter when the cold makes
them hungry and the tourists from whom they have been levying
tribute all through the summer have disappeared. In cold weather
they sit all round the Piazza keeping a greedy eye on the clock.
They hunch up their shoulders against the cold and fluff out their
breeches until they seem to be clad in pale grey knickerbockers
and become quite round like fledgling sparrows. When they see
the pigeon man come into the Piazza with his bins of maize there
are uneasy sporadic flights, small platoons make false starts, hungry
individuals make vertical take-offs and flop back with a disappointed gr-gr-gr on to their ledges. At the first stroke of the
clock the *préposé au maïs*, as the French would probably call him,
scatters his maize in a great winding stream across the Piazza and
ten thousand pigeons launch themselves with a tremendous flap-
ping of wings to gobble and jostle in the great square while late-
comers anxiously wing in from all parts of the town. This is the
only time that the pigeons are at all likeable. They are really
horrible birds. They trade on their position as all privileged
classes do. They are dirty, they are a nuisance and they do great
damage to the buildings of Venice. Being privileged they have no
intention of reforming but create progeny for all they are worth to
carry on the despotism with which they rule. The Venetians, who
until recently have been putting up with them almost without
complaint, are now beginning to resent them. Hitherto there has
not been much evidence of the damage that they do, only of the inconvenience from pigeon droppings to householders and passers-by.

Since the 1966 flood in Venice a survey has been made of the state of many buildings and the condition of the stone. Evidence is now available. Apart from the visible filthiness of statues and the façades of buildings and the stains made by the acid in drop-pings on stone and marble, pigeon guano in a layer four and more inches deep is causing the ceiling of the deconsecrated church of Santo Spirito on the Zattere to buckle. It was also discovered that the pigeons had deposited over half a hundredweight of guano on the back of the huge ceiling painting by Nicolo Bambini in the church of S. Moisé. The city health authorities are now consider-ing how to reduce their numbers to protect the people and the buildings. Since their nuisance value derives chiefly from their philoprogenitive ways there is a sort of poetic justice in the remedy which the Venetian health department is trying out. It is feeding them with contraceptive pills.

During this rather idle time I spent entire weekends looking at the pigeons, 'beholding things', or discovering 'little terraces to lean over'. I am left with a kind of kaleidoscope of things seen and heard. I hung about the Piazza waiting for the platoon of bersa-glieri to come at their strange loping gait from the Molo where they disembark to attend the ceremony of striking the flags of Italy and St. Mark on Sunday evenings; a fine symbolic ceremony. I watched funerals with the evident human sorrow made somehow more bearable by the indifference of the gulls swooping overhead, the not-to-be-repressed cheerfulness of the little waves slapping at the barge with a shower of bright splashes and the gondoliers' solemn suits of black jauntily set off by their piratical hats. I found strange saints to decipher like S. Agiopo (S. Giobbo) and S. Trovaso (SS. Gervasio e Protasio) and some who refused to be deciphered like S. Bruson. I found that I should have to pay my telephone bill in the gothic cloister beside the church of S. Salva-tore. I was invited to a meal at Easter and thought the Venetians had got their symbols a little mixed when they ate the paschal lamb. I had time to join the football crowd returning from a match at S. Elena. Good-humoured, stripped of the week's irritations by the strange catharsis of the game, they came in genial little groups in a long black agitated sea-serpent undulating up and down the bridges all along the Riva degli Schiavoni. The crowd made a noise like a gentle tide running up a pebble beach with the firm attack of soft shoes up the bridges, the little resonant echoes below the arches and the quick trickling away of footsteps down again to the levels of the Riva.

Mary McCarthy says somewhere that nobody discovers an original perception about Venice. She is quite right. Somewhere else she notes (a little to my chagrin for I really thought I was the first to notice this) the quality of the Venetian walk. I learnt about it from the football crowd and also from the late-night passer-by (Venice is never entirely without the sound of footsteps) under my window by a bridge on the Riva degli Schiavoni. One could hear him coming from a distance and could pace him along the Riva. He usually came with a quick, light rhythm slowing and lengthening for a few paces to mount a bridge (now he had reached the Ponte dell' Arsenale, and now S. Martino and now the Pietà) and compensating by a rapid tap, tap, tap down the other side to resume the quick, light pace along the level *calle*. I practised the technique of walking down the broad steps of bridges which consists in simply transferring one's weight a little forward, automatically bringing one's legs into rapid action to keep one in balance and carrying one quickly down to the level of the lane. I learnt to weave rapidly along the narrow streets without dodging inelegantly between groups or jostling slower traffic. I even learnt the rudiments of umbrella management in narrow crowded spaces. Walking became one of the many simple pleasures afforded by this complicated recondite town.

I found, at this time, that my attitude towards Venice was changing. This was partly because I was beginning to know the town and that I no longer had the tourists' *optique*, that quality of vision composed of a ready appreciation of the externals of someone else's condition. But it was chiefly that I had become involved with Venice in the same way that I was involved with my family or England or France. So I became impatient with Venice when it appeared at less than its best and critical of its imperfections as

I was impatient with my poor family when they did or said silly things (which is unjust because they are sensitive and intelligent people); or with England when it is small-minded and illiberal and preoccupied with keeping up with its insular Joneses (when I know that its imagination and political flair are unimpaired); or with France teetering on the brink of anarchy like a pirate's prisoner blindfold at the end of the plank (when I know that it will react with honour, sense and dignity, and even with panache, at the eleventh hour or a little later). So with Venice. It was provincial, it was misguided, it could often not tell the wood from the trees, it was maddening to shop in and absurd about its future. But all these things, as well as the splendour and the tarnished glory and the marvellous light, made up Venice. In fact, I found I loved the place.

CHAPTER V

REPLANNING THE VINEYARD

ONE of the reasons for the delay in starting to build the house was
the lack of fresh water for mixing the mortar. Torcello has always
lived on its wells. Some, like my neighbour Giannino, commonly known to us as *il papa di Rosetta*, were lucky and had
sweet water, but mostly the water was brackish in taste and stained
brown from the rust in the well pipe. At first I had thought it
would be necessary to sink a well and had gone so far as to hire
a water diviner (this being the approved way to do things in these

parts) who stepped about briskly amongst the weeds holding his
forked divining rod at arm's length before him. The water table
is very near the surface in Torcello and his rod flipped about in
an undisciplined way in his hands. He finally marked down for
me the best place to sink a well. There were several things against
this spot. It was very inconveniently placed in relation to such
planning of the vineyard as I had so far done. It was also close
to the existing shallow well which provided a cloudy, heavily
brackish liquid which we had already decided was unsuitable for
mixing cement; so near, in fact, as to make one doubt the quality
of the water that a deeper well might produce.

For these reasons it seemed better to delay a decision about
sinking a well until we knew more about the new pipeline
designed to supply the growing tourist communities along the
coastline of the Cavallino *lido* which protects the lagoon from the
Adriatic Sea on the other side of the Porto S. Nicolo opposite
Venice. This gave a certain urgency to the project and I was
delighted to discover that work had already started on it. Even
better was the news that it was to run through Torcello and Burano
on its way to Cavallino. In the event it ran straight through the
middle of my land.

The pipeline brings us a light, sparkling water from the foot-
hills of the Alps behind Treviso. In taste and purity it matches the
best I have tasted in Italy where the Romans have left a strong
tradition in the quality of drinking water. This is unfortunately
being impaired, as in most other industrial countries, by the huge
supplies demanded by factories and a greatly increased population.
When I was a child in Rome we used to have our drinking water
brought in from a favourite spring and this is perhaps what
started a lifelong interest in the taste and quality of water. I may
forget the details of monumental meals eaten here and there, a
thing my French friends would never do, but I can usually tell
you what the water was like. In too many English towns we are
approaching a dead level of chlorinated municipal cleanliness as
though all the drinking water in England were pooled and then
shared out again. It is, no doubt, not harmful but it is now hardly

ever a pleasure to drink. Among University towns Oxford scores particularly low marks for a water which offends the nose, ruins the palate and weighs down the digestion. What effect, one wonders, can it have on the mental processes of the denizens of that great university? Paris does rather better, in variety at least. Some districts (and it seems to me only just that they tend to be the *beaux quartiers*) are supplied with such a lifeless hygienic water that it cannot only be for reasons of health or snobbery that the inhabi/ tants drink water out of bottles, choosing between the natural bubble of some kinds and the fabricated fizz of others and between the somewhat dreary purity of Evian and the faintly oily heaviness of Vittel. Other districts still draw their water from ancient water supplies which used to serve the old city. In the part of Paris where I live there are two ancient supplies. One is in the present rue Henri Barbusse, earlier called the rue d'Enfer (a nice folk/memory of Via Inferior), where a reconstructed eighteenth/century supply still exists. The other is an old reservoir in the convent of Les Filles du Bon Pasteur near the top of the Boulevard Raspail. It was built to supply the Luxembourg palace and its gardens and the surrounding quarter and drew its water from Rungis to the south of Paris. Now that the suburbs are spreading fast on that side and the Paris *halles* are installed there this supply is presum/ ably augmented from other sources. I certainly get water which is good enough to put in the refrigerator to chill.

Since the pipeline is intended to improve the water supply for a developing tourist area it is only by a fortunate chance that I have access to it. Only one outlet is provided for Torcello and by another stroke of good fortune this is conveniently situated on the towpath by the canal bisecting the island, just on the other side of my neighbour's land. I struck a bargain with him by which he let me bring the water across his land provided that I laid a branch pipe to the outside wall of his house. In this way we both benefit from this good sweet water. We are the only two houses in Torcello to do so. All the rest continue to use wells. The new pipeline also supplies the neighbouring island of Burano which has a growing population of workers who commute to jobs in Venice and

Murano, but a similar community in Mazzorbo, which lies just off the route of the pipeline from Torcello to Burano, cannot draw upon it. The water in Mazzorbo is unpleasant and there was very great ill-feeling and much painting of abusive slogans on walls when the inhabitants discovered that they were to be left out of this improvement in a basic need. I do not know whether the water is simply brackish in a particularly nasty way or whether it is, in fact, polluted, but the people of Mazzorbo are certainly convinced that it is bad for their health: so much so that at Maddalena's little restaurant they refuse to let you drink it, automatically bringing bottled water to the table.

We should welcome the current preoccupation with pollution if it can bring better conditions in a world in which a rapidly increasing population is crowding into huge urban communities. What is strange is that pollution should be thought of as something new. Queen Elizabeth's progresses, which proved so disastrously expensive to some of the great lords who provided her and her vast court with hospitality, were not only the consequence of policy but were to some extent imposed by the fact that the palace was so polluted as to be a danger to health after a residence of several months and had to be abandoned until the filth and stench evaporated. The château at Versailles with a population of something approaching thirty thousand at its peak and with only the most primitive sanitary arrangements must have been a nightmare of pollution. Even in the 1890's the river at Cambridge was a sewer. In that enchanting book on her childhood, *Period Piece*, Gwen Raverat says, 'There is the tale of Queen Victoria being shown over Trinity by the Master, Dr. Whewell, and saying, as she looked down over the bridge: "What are all those pieces of paper floating down the river?" To which, with great presence of mind, he replied: "Those, ma'am, are notices that bathing is forbidden." ' As Mrs Raverat seems to have spent a great deal of her childhood in and out of boats, it is perhaps as well that she lived upstream of the town.

While I was well content to have the pipeline run through my land, there was a reverse side to the medal. The Venetian municipality decided to take up its option to clear a belt fifty metres wide through the middle of the vineyard in order to lay the pipe. This was to leave a ruinous scar like those swathes cut out of the Viet-nam jungle for the purposes of war. Yet it was not wholly destruc-tive, partly because it caused me to make up my mind about the future of the vineyard and partly because it produced 800,000 *lire* in compensation calculated on the estimated lifetime product of the trees and vines which were to be destroyed. The compensation did not, of course, come at once and, indeed, the sum itself was the result of long negotiation, but at last it was paid and happily coincided with some of the larger bills presented by the builder.

The more I thought about the aged and neglected trees and vines that were left, the more I pondered over the cost, in wages and upkeep, of maintaining the vineyard as against the likely yield, the more I counted the cost of putting the destroyed fifty metre section into good heart again, the more doubtful I became about the economics of keeping the vineyard going in its then state.

It had been planted in the local traditional pattern of vines set in squares held up on Y-shaped supports with fruit trees dotted about haphazard in each square. The effect was of an overgrown chequerboard. The trees and vines were closely intermingled and could not breathe and develop and the shade they cast hindered the growth of the peas and beans and other catch crops planted amongst them. The pattern was expensive in labour as no tool bigger than a hoe could be used in such a cluttered place. And then too much valuable space was taken up by the paths round and between each square. It was charming, it was romantic, but one had to concede that it could never earn its keep.

Clearly a radical reorganization was necessary. But this was not without its problems either. In the first place the cost of uprooting trees and vines, of ploughing the whole site, of filling in the shallow well, putting down paths in a new pattern and replanting with good stock would cost a good deal. I was not wanting in friends who kindly pointed out to me that if I was getting 800,000 *lire* (about £530) compensation for the produce of a swathe fifty metres wide I was, in effect, intending to plough up a whole patrimony. They were, of course, quite right. Then I should have to support the vineyard for at least five years before I saw any financial return, for the fruit trees would only then be beginning to provide a marketable crop and vines do not reach their full production until they are between five and ten years old. Could I carry the cost of wages, implements, fertilisers and all the hundred and one other things required on a holding of this size for so long? On the other hand if I left the vineyard as it was it would be a totally uneconomic business proposition by the time I retired and I would then have no money to retrieve the situation. Whereas I had the money now and if I spent it (the word 'invested' seemed to me, even then, to be going rather far) on the vineyard it would, at worst, be paying its way and, at best, adding to my pension when I did retire. And anyway, let me be honest about this, this is what I wanted to do. So I did it.

The first thing to do seemed to be to consult an agricultural expert. I am always for consulting experts though not always for taking their advice. I, too, am an expert, perhaps that is how I know. He was a pleasant man and we spent an agreeable morning walking about the vineyard and deciding what ought to be done. The first thing, he insisted, was to fence the whole place. I did not want to do this partly because it seemed to me to be a waste of money but chiefly because I liked the informality of never being quite sure (except when putting in crops or trees, when one made sure) as to whether one was on one's own land or someone else's. But he told me that according to Italian law pilferers could only be prosecuted if they had penetrated into enclosed land. Unfortunately, a good deal of pilfering of crops occurs in Torcello though whether this is, as one might say, for professional reasons, or whether it is visitors to the island (local as well as foreign) getting out of hand it is difficult to say. The estimate of the cost of the reorganization (including the horrible fence) calculated by the agricultural expert came to two million *lire* (about £1,330). One day, when I was in Paris digesting this disagreeable news, and coming to the conclusion that it was absurdly high, I had a telephone call from my man of affairs in Venice, asking me if I had agreed to the cost because the expert and a team of men had already begun work. We now learnt that the expert was his own contractor. My reaction was immediate: to refuse the estimate and stop the work.

So, two years after buying the land, I was without an architect for the house and without an expert for the vineyard. I was really getting on.

That was in March 1965. On May 8 of that year I was out at Torcello when a rather slight grey-haired man, with a modest demeanour, emerged from the undergrowth and demurely picked his way to where I was having a picnic under a broken-down old fig-tree. He had come, he said, at the suggestion of my man of affairs who thought we might like to have a little talk together.

'Oh, yes?' I said, not very interested.

He was, he said, the clerk of works at St. Mark's, in charge of the fabric of the basilica.

'Oh, yes!' I said, much more interested.

Did I know the basilica at all?

Indeed, I did. But only an outsider's view, of course.

Mmm . . . Perhaps I would like to visit it with him sometime? Just then, he was having trouble with the horses. The horses had a disease. The bronze was being pitted by some acid in the atmosphere. Very troubling. I must come and see.

Later, after we had adjourned to a bar and I was drinking coffee and he was drinking one of the rather sticky drinks I found he was addicted to, he told me that he was also having trouble with St. Mark. The new liturgy required the altar to be moved forward to allow the priest room to officiate between it and the Pala d'Oro, facing his congregation, and St. Mark, in his stone tomb below the altar, had to be moved forward too. And he weighed tons. Ten men with crowbars had not been able to shift him. He did not know how he was going to move him, he gently mused.

And so *de fil en aiguille* he had come to offer to help me with the *sistemazione* of the vineyard. It would be diverting, he thought, to work out what ought to be done. In this way I acquired, in Emilio Fioretti, a good friend and excellent adviser.

Barzini, in his book on the Italians, says that it is the dream of most Italians to carry through a *sistemazione*. I did not really believe him until I started to reorganize the vineyard. Then I found that everyone I knew was enthralled by the idea of reorganization *per se*, by the pure intellectual challenge of finding a solution. I was delighted with this attitude since it brought me new friends and much advice. Some of the advice would have led straight to disaster but it was given with such kindly intentions and was the fruit of either a sudden brainwave or, more touchingly, of prolonged cogitation over someone else's problems. It was sometimes difficult to avoid suggestions without giving offence. We always had a band of interested observers ready to give a hand, but more especially advice. It was these people quite outside the situation (like the postwoman who had just come to bring a telegram, or the man delivering a load of bricks or his friend who had just come along for the ride) who were particularly tenacious about the excellence of the solution they had found.

Fioretti threw himself wholeheartedly into the problem. He spared neither time, trouble nor ingenuity and the plans finally adopted were largely his. He was constantly on the go, seeking advice here, telephoning for help there, asking for estimates and price lists. When we started work he brought his own team of men from St. Mark's who, on several consecutive Sundays, fenced the land, built the barn and put up the stakes and wires on which the vines were later to be trained. Throughout I had the impression that he was having tremendous fun. I certainly was. Looking back on that time, I see that we made some mistakes. We should, for instance, have made the contract ploughing men clean out the ditches which drain this low-lying land. It would have been simple and inexpensive to do on an empty site and one day soon it will be a complicated and costly exercise that I shall have, willy nilly, to undertake. We simply did not happen to think of it at the time. The rows of vines, I now realize, are too far apart. We could have got in another two rows and consequently two more rows of artichokes between them. This, I think, happened because of Fioretti's enthusiasm for an elegant mathematical solution of twenty rows five metres apart: a *sistemazione* readily lends itself to this kind of enthusiasm. It was also a mistake to plant four different kinds of vine. This happened probably because we both suc/cumbed to a romantic view of my cellar stocked with several kinds of wine made on the estate: château/bottled in fact. The truth is that the local wine is poor. It makes quite pleasant drinking when it is new, then being a sort of *vino santo*. None of the mistakes was disastrous and they were a small price to pay for the experience of living through a *sistemazione* with all that this involves of a baroque bravura performance, of passion, of hard work and the application of common sense. For the intellectual pleasure, as I very clearly saw in Fioretti, derives ultimately from finding a neat and work/able solution.

I had three objectives in reorganizing the vineyard. First, to re/plant it with the best strains of vines and fruit trees and to prune and maintain them properly from the start. Second, to plan it to econo/mize time and effort so that one man could work it with casual help at harvest time. Third, to select crops which would provide a reasonable return for the labour and money expended on them.

We decided to divide the vineyard into four large and two small sections arranged on either side of the central walk running from the inland gate to the landing stage near the house. This permitted the replanting in rows running east–west the length of the vineyard so that all the vines get the maximum sun at all levels on each side of the tall, straight supports. The rows were set five metres apart to leave easy working space for the tractor and for catch crops (mostly tomatoes, aubergines, *zucchini* and melons) to be planted in between. The two small sections near the gate were to be devoted to artichokes. Fioretti and I calculated that we could do all the work required for less than the previous estimate of two million *lire*. In the event we also spent two million *lire* but this figure included a better fence (since I had to have a fence) and the cost of building a large and commodious barn.

While we were planning the *sistemazione* of the vineyard I came to know the basilica of St. Mark from a different point of view from the tourist's or the art historian's. I used to go there to discuss one point or another with Fioretti and he used to take me about with him on his rounds of the work in hand. St. Mark's is the centre of his life and he has an intimacy with the structure that even the architect *procuratori*, who are in charge of the building, do not have. Fioretti goes about the basilica with all his senses alert, with all his antennae receiving messages. He tends to have a slightly preoccupied look and however much he may be listening to you, he is really listening to the building. He will suddenly stop in his tracks, go back a few paces searching the uneven floor and stoop to inspect a patch: walking across the marble inlay he had felt something that was not quite right. He will run his hand over a section of wall as we go by and will mutter about an infinitesimal increase of dampness. He will rap a pillar as he passes and say 'Tcha' disgustedly and invite me to rap it too and listen to the hollow ring where the rubble has settled over the centuries leaving a ghost pillar faced with marble. To comfort him I tell him that the pillars of English cathedrals are also often filled with rubble, but it is nothing to him if our cathedrals are badly built. His basili-ca should be perfect. He knows the basilica so well that even along the rough-hewn floors of the catwalks linking the pillars high above the nave, he would point out obstacles to me but never need to look where he was putting his own feet. Sometimes I used to find him in the upstairs workshop making sure that the patching of the mosaics taken down for repairs was proceeding well, or up a ladder supervising their replacement on the walls and vaulted ceilings, or in the crypt draining away the flood water sloshing about the tombs of the patriarchs. Sometimes I would wait for him in the sun of the Piazzetta till he came down from visiting the angel on the tip of the Campanile, or in the cool of the nave sitting on the pedestal of a column as uncomfortable as any medieval monk perched on a miserere, and then I would be hailed in a suitably muted tone ('Eh! Signorina!') and find a friend from Fioretti's Sunday vineyard team grinning at me from below a mason's paper hat precariously at work preventing St. Mark's (jerry-built and glorious) from falling down.

*   *   *

One day in the late summer of 1965 an enormous tractor arrived
on a pontoon with all its grabs and ploughshares arranged about
it and lumbered over a plank bridge into the vineyard. Its arrival
had to be delayed for two days because the man who ran the firm
with his two sons had had the tractor upset on to him and had
been killed. Rather silent, the elder son, now the head man in the
firm, came to greet me, apologizing with grave courtesy for the
delay. Then with his brother and the new man they had taken on
they started manoeuvring the huge, murderous tractor to start
work. And again I wondered what had given rise to Northern
European views of the Italian as shallow and flighty. In spite of
the laughter and the gaiety there is something almost dour about
the people of the Mediterranean. The working people of Italy live
an immeasurably harder life than ours and react with a dogged,
fatalistic courage.

First the driver fitted a grab and seized a tree. He backed off with
a hideous puttering and, enveloped in a blue haze of paraffin
fumes, tugged and wrenched till the tree came free. He then
crashed off amongst the vines to deposit it on a dump he was
making by the lagoon. When all the trees in the vineyard had been
uprooted, he went round with the grab biting and snatching at
vines and tossing them aside to be cleared away by the other two
men who followed him round.

The next day the driver attached a huge two-foot ploughshare
to the tractor and went up and down turning back the earth in
silky brown swags. It is a beautiful light rich soil that will grow
anything and, in its fertility, reminded me of the celery beds in the
Isle of Ely. But in one section brick rubble was mixed in the soil.
We went over to look and saw that the edge of the rubble seemed
to follow straight lines which met at a right angle at the lagoon
end. At the other end the rubble-strewn part seemed to be following a curve. As the tractor continued its journeys up and down
it gradually came to us that it was outlining in the moist newly-
turned earth the ground plan of a church. There was no point of
vantage on this flat land from which to look down on it except
the tractor. We all clambered on top of it for a better look. It was

the vanished church of S. Tommaso dei Borgognoni. It had served the great Cistercian monastery which owned all of this part of the island, still known as dei Borgognoni. As we looked the ground plan began to fade as the soil dried.

The men looked pleased and a little awed and helped me down silently, offering, as always, a wrist rather than a dirty hand to steady me, and we all went home quietly because of the unexpectedness of it.

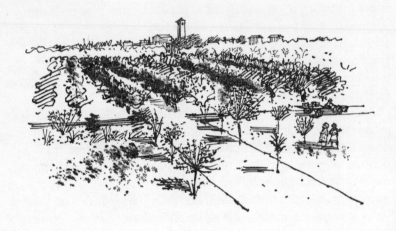

CHAPTER VI

A NEW ARCHITECT

THE work in the vineyard had deflected my attention from the frustration arising from the Soprintendenza's rejection of the sketch plans. The fact that the well-water was too brackish to mix the cement and that, consequently, building would have to wait until the new pipe-line had been connected was, in a way, a relief. I could, at least, tell myself that it was not only my incapacity to find an architect which led to all these maddening delays.

Early in 1965 two fortunate things happened. First, my man

of affairs suddenly revealed himself almost as thwarted as I was by
the Soprintendenza's rejection of the sketch plans. Until then,
while being very helpful about the complexities of buying the
land, he had seemed only remotely concerned about the building
of the house. He now began to take an active interest and to talk
about 'our' project. One day in March he suggested that we
should go and see an architect called Marino Meo. He told me
that, besides being widely known in Italy, he was one of the
architect *procuratori* who advise on the preservation of St. Mark's
basilica. He had built a number of houses in the lagoon and had
very good relations with the Soprintendenza. All this sounded
promising but nevertheless I was very reserved and wary as I
followed my man of affairs up the dark, cold Venetian stairs to the
top of a house near Rialto. The top floor was lit by a big skylight
and, as we waited on a newly laid marble floor to be let in, I
admired the design of the fine, wrought-iron glass-backed doors
which led into his offices. We were led into a corner room and a
man got up from behind a drawing table littered with plans and
flanked by a spanking new calculating machine to welcome us.
My first impression of him was of a gentle person whose feet
seemed to be hurting him. Then I noticed what a comfortable,
practical and handsome room he sat in and realized the skill that
had gone into creating offices like these in the framework of an
ancient Venetian house. I looked out of the windows at the surrounding roofs, clothed in mellowed tiles and progressing, ridge
after ridge, to a *campanile* in the corner. I looked at the photographs
and plans pinned to the walls—good buildings, clean and un-
cluttered—and found myself observed by intelligent eyes and a
faint ironic smile. He seemed to find me funny and that was a good
start.

We talked about my ruined cottage and the Soprintendenza's
verbal agreement to allow me 'to rebuild and slightly enlarge' it.
We talked of the direction it faced and its relation to the lagoon,
and about the sort of house I wanted and the amount and kind of
accommodation I needed. He then took a sheet of drawing paper
which he put before me across the table and, upside down from

where he sat, he drew first a plan of the ground and first floors and then an elevation with quick, evocative strokes of a soft, black pencil. I looked at him in astonishment.

'That's exactly the sort of thing I want,' I exclaimed. 'When I come back next month can I see the sketch plans?'

'Next month! Come tomorrow.'

'But tomorrow is Sunday.'

'Well?'

My spirits rose. I had, I thought, possibly found an architect and perhaps, one day, I would have a house.

Looking back on those months of working with him, and particularly those first few interviews, I became even more per-plexed about the relationship between an architect and his client. One of the difficulties of finding an architect, or indeed any other professional man, is that you cannot interview him and go away if you do not like what you see. You have to trust your teeth to a new dentist and your ailments to a new doctor. And yet that first meeting with Marino Meo was in a way an interview: a two-way interview. We both knew that we were summing up the other.

I thought later that an important element in an architect's equipment, when first approaching a client, must be the capacity to draw. Certainly that rapid, vigorous little drawing materially helped to break down my suspicious reserve. I realized later that I had also been taken by the technical virtuosity of doing a drawing upside down. I cannot draw and this little bit of panache appealed to me. Was it calculated? I do not know. But the whole of this little episode told me something about the man which I rather liked. It must also be very helpful to the architect to be quick in seizing on the essential of what a client wants. This is a much more complicated requirement because many clients must have only confused and contradictory notions of what they want and many others must be inarticulate and unable to convey require-ments which are clear enough to them. I suspect that most of us have certain details of what we want in a house quite clear in our minds but that the whole is rather indeterminate, even misty, like faces seen in a dream. The speed with which Marino Meo sifted out what were, to me, the essential points about the house certainly helped to create confidence and I was prepared to look on his further suggestions with a friendly, rather than a suspicious, eye.

There must be a large part of psychology in an architect's equipment, though with Marino Meo I have never been sure whether he is particularly diplomatic and skilful in dealing with clients or whether his success comes simply from being himself without self-consciousness. I found him, almost at once, congenial company and we very soon seemed to be talking on a whole wide range of subjects which had nothing to do with the house. We had plenty of opportunity to talk on the thirty-five minute journey from Venice to Torcello and it was on board the *vaporetto*, with plans and estimates and illustrated brochures of baths and heating equipment spread over the surrounding seats, that much of the detailed planning was done. In between times he pointed out a fisherman's technique, or an interesting building, or told me the best place to buy gramophone records or took me on a detour to look at a picture. He taught me more Italian than almost anybody else by refusing to speak French (I am sure he can) and looking extremely pained when I did dreadful things to his beautiful language.

Because distances are so great and transport in the lagoon so slow, the planning and supervising of the house took a disproportionate amount of his time and he kept in touch with his office and his other jobs by constant telephone calls. We were continually popping into bars to telephone and I came to know a wide variety of drinks because I very soon decided that I could not face the strong sweet solid coffee at the frequent intervals that the Italian digestion appears to be able to manage. For the same reason we had many lunches together and if I know now where the best fish is to be obtained (it is at Achille's restaurant at Cavallino) or the best wine or *pasta* between Punta Sabbioni and Jesolo it is as a by-product of building a house.

Besides all this an architect must be seen to be a practical man and a good manager: virtues hard to fuse with the contradictory role of a magician making substantial his client's dreams. Most

clients, I imagine, are anxious about money. Few realize how
much building costs, are shocked that the minor changes they
suggest should prove so expensive or that the final cost seems
inevitably to be a good deal more than the estimates accepted at the
beginning. A great many must, like me, be operating on the
fringe of their financial limits, when not beyond. All this creates a
state of anxiety which would speedily lead to a desperate situation
if one did not have confidence in one's architect's good sense and
practical handling of problems.

I have been very fortunate in all these respects so that when
later, some difficulties arose between my architect and myself, they
never affected my appreciation of his personal or professional
qualities. I hope he can say as much for me.

He soon saw that I was interested in buildings and showed me
work which he, and other architects, were doing. This was how I
came to know the new little resort planned at Lio Grando. There
a hotel, restaurants, a neatly laid-out estate and a yacht marina
have been planned near to Punta Sabbioni between the main road
which runs like a backbone along the middle of the Cavallino *lido*
and the coast of the lagoon. It is a pleasing contrast to the tourist
sprawl along the rest of the coast. Here Marino Meo has created
a charming waterfront, inspired by the gay painted cottages of
Burano, yet entirely in the modern idiom, strung along the beach
and the marina. Though I am interested in buildings he was
quick to see that I did not have enough technical knowledge to
know the effect of various building techniques. So he showed me
examples. Thus, when we went to Cavallino to have lunch, and
to telephone, he showed me, on the wall of the town hall, exactly
the texture and the colour of the stucco which he had suggested
for the exterior of the house. We walked up and down the town
hall running our hands over the surface here and there until we
found a patch which had the colour and the degree of roughness
which seemed to us just right for a house in the country. Then we
called over the builder who was standing by himself a little way
off, rather ashamed of the exhibition we were making of ourselves
and pretending he was not with us, and we said 'Do that'—and

he did and it has proved a great success. In the same way he took me to Padua, where he had to go for other reasons, to select the baths and washbasins and taps, and invited me to his house in Asolo to see the tiles in his bathrooms. If I had had any lingering doubts about the quality of his work and his taste, one look at his own house would have dissipated them. Intensely Italian, even Venetian, which is right, in atmosphere, very handsome, practical and comfortable yet ingenious in using the slope of the land and elegant in design, it is a house which is relaxed and friendly and beautiful and I know very well that this atmosphere is a reflection of the personalities and taste of both Marino Meo and his wife.

I found these months of planning and building the house exhilarating and absorbing in spite of the accompanying anxieties. I never for a moment doubted that my architect would build me a beautiful and convenient house. I now know some of the elements to look for in one's association with an architect. I still do not know how you can tell in advance if he has them so I still do not know how to choose an architect. The basic principle is clear and as expressed by Jane Drew is 'You need something of quality in the first place'. I do not know how you ensure that you get it. Perhaps it is luck. If so, I am very fortunate because I certainly got it.

The second fortunate event was to be given, by a neighbour in Torcello, a copy of an eighteenth-century print showing the great church of S. Tommaso dei Borgognoni whose ground plan we had just revealed in the vineyard, and showing also, to the left of the print and on the site of my ruined cottage, a small, neat villa with a walled garden. Of the villa, the walls and the trees, nothing unfortunately remains except the rounded stone course running along the wall of the waterfront. This still runs right along my land. On the other side of the church my neighbour's house, now with dilapidated farm buildings replacing the retaining wall which has disappeared, still stands beside the canal some hundred yards to the left as you land from the *vaporetto* at Torcello. The canal was considerably wider then than it is now, because the *barene* (those sedgy flats covered with coarse grass and hazy with blue sea-lavender in the autumn) have encroached on the deep canal. It is still a main highway, however, and daily huge barges laden with corn laboriously chug past my sitting-room windows on their way to the silos at Treviso. The fishing boats and the *sandali*, those elegant black country counterparts of the gondolas, have not changed at all. Apart from the satisfaction of possessing the print and of knowing what the house looked like two hundred years before, it proved of unexpected practical value at a moment when we were re-opening negotiations with the Soprintendenza.

My second visit to the Soprintendenza was in marked contrast to the first. I now accompanied a man who was welcomed as a friend and a respected colleague. The conversation wandered over various projects of interest to both sides and relating to the preser-vation of old buildings and the building of new ones in old parts of the town. Marino Meo then unrolled his plans and even they seemed to have a more professional air, a look of knowing what they were up to and of being confident of their reception. In spite of their benevolent aspect the officials of the Soprintendenza were very cautious, as indeed they should be. They made some criti-cisms. They did not like an overhang that had been planned over the lagoon. They thought that the balcony at a window situated over the front door would give too sophisticated an air to the landward façade. But these were minor criticisms and, after a decent hesitation and a whispered consultation, we conceded them. Then the main doubt in the Soprintendenza's mind emer-ged: the size. The new design was, they said, charming and fitted perfectly into the surroundings, was entirely *lagunare*, in fact, recall-ing many lagoon farmhouses, but it was so very much larger . . . Their doubts hung in the air and I perched anxiously on the edge of my chair. Marino Meo almost shyly played the ace card. This print, he said casually, which we had been given, showed the house as it was originally planned . . . unfortunately only the foundations remained which he proposed to strengthen and use for the new house . . . and here, on the print, was the eighteenth-century house which still existed next-door . . . the proportions between the two houses in the print had been maintained in the new design ... Objections seemed to melt away before the sweet reason of his approach. I must say I went away rather cock-a-hoop. Final plans had, of course, to be submitted for scrutiny but the precious agreement in principle had been at last obtained and who cared about the overhang or the balcony over the door anyway? In fact, of course, I do regret the balcony every time I search for the keyhole in the pelting dark of a lagoon night with the sea water slopping against the wall scarcely a yard away. Then I hate myself a little because I forgot to make the point to the Soprinten-denza that the balcony was no mere echo of Romeo and Juliet but intended to provide protection from the weather for those seeking to get into the house.

The print was useful in other ways too. One day we received a severe note from the Venice municipality to say that it had come to their ears that we were proposing to build a house on the actual edge of the lagoon. It had perhaps escaped our attention, it went on, that regulations existed regarding the building line which had been fixed at two metres back from the water's edge. We put the print of the house and some photographs showing its present state into a nice clean envelope and set off to the *palazzo* on the Grand Canal which houses the *Comune*. There we eschewed the lift. It is the only one I know of in Venice outside the grander hotels and rises majestically, if a little erratically, like a medium-sized winter garden emptied of its palm trees and ferns. We displayed the photographs of the ruined cottage. There it was on the very brink of the lagoon and, as its state clearly showed, had been there a very long time. But, they replied, that ruin was only a shed, a place to keep potatoes away from the frost. We, it appeared, were intending to build a house, which was quite a different kettle of fish. Again we explained that this sad ruin was all that remained of a house which had been lived in for many generations. This print showed its original state. And this official too was delighted with the print and since what existed existed, he initialled the plans approving the rebuilding of the house on the original site. This is how it happens that I can go fishing from my sitting-room windows.

The print also gave me a first vivid glimpse of what Torcello had been like in the past. It is hard to believe now that it was once a prosperous well-populated island. Because of its remoteness Torcello retained a certain political independence even at the height of Venice's power. It was the seat of an autonomous *podestà*, supported by a Council and judiciary and its own system of nobility, all, of course, very closely modelled on the practice of its great neighbour. At one time it even had the neighbouring island of Burano as a dependence. Now the latter is thickly built over and, on the rare empty spaces, new flats and houses are going up to accommodate the increasing population. There is an urban bustle about Burano in spite of the fishermen sitting on the quays with their nets stretched out over their feet and their hands and toes pulling the net taut like embroidery over a frame, as they work over it looking for holes, or the women sitting in their doorways tirelessly throwing the little bobbins to and fro as they make the lace for which the island is renowned. These are quiet and sedentary occupations yet requiring constant activity and constant attention. Torcello has none of this air of activity. Our fishermen mend their nets sitting in their own boats moored by their own houses up the forgotten creeks of the island, and while nearly every woman on the island has been trained in one or two of the simpler stitches of lace-making few of them are reduced to accepting such poorly paid employment. For most, it is not well paid. A great deal of the work of building up a piece of lace demands only the use of simple stitches. This done, the little cushion with its bobbins is passed to a more skilled and better paid worker until at last it arrives in the hands of an expert lace-maker who gives the final touches to the *punto buranese* which has made Burano famous since the sixteenth century.

In the lace-making school on the *piazza* of Burano you can see little girls learning how to flick the bobbins in and out or practising the more complex stitches of the exquisite embroidery which they also make under the benevolent but reasonably firm guidance of a nun. The work of the school and that done in the many cottages of Burano may be inspected and purchased at the school. It is not cheap, but why should it be? It is very fine. It is skilled handwork. It takes many years to become skilled in this trade. At one time in the last century the skills were almost forgotten and the craft was revived, thanks to the generosity and interest of Paulo Fambri and of the Countess Adriana Marcello who saw to it that the one remaining old woman who knew the secrets of making the *punto buranese* should pass it on to younger women. At that time Burano had become a very poor, fever-ridden community of fisher people and, apart from saving a dying skill, the importance of the initiative was to bring back some prosperity to the island. Today these islands still need people like Paulo Fambri and the Countess Marcello, to restore old industries and to create new ones to enliven and strengthen the life of the lagoon.

Our occupations in Torcello are more simple and solitary and when we get home we tend to go in and shut the door and live our lives out of sight. The main meeting place is the *vaporetto* stop and the path beside the canal which leads from it to the basilica. The *vaporetto* stop serves the same purpose as the fountain in other villages. It is convenient, it is shady and we all have to go there. It is true that we have a fountain near the *piazza* with good fresh water gushing into the nearby canal. but it is not a meeting place except for the boatmen waiting to row their customers back to the *vaporetto* stop in their *sandali*. There are no shops on Torcello (you have to take a sea voyage to Burano to buy even a loaf of bread), though there is a very famous restaurant (Cipriani's by the *piazza*) and a little bar where you can get a drink while you wait for your turn to telephone. We all have to take the *vaporetto*. The men to go to their work and the children to school and the women to shop and visit their relatives in Burano. Most of the twenty-six houses on the island are strung along the canal and as the time for the hourly *vaporetto* approaches the doors of the quiet, apparently abandoned houses briskly open and people emerge to join the little groups hurrying intently along in the sun, exchanging news and views, and even snatching a moment to glean a sidelight on the day's events from those luckier individuals already returning from Burano laden with gossip and groceries. So we are very well informed about our own affairs. Certainly never more than an hour elapses between my arrival home and the first visitor coming to transact business, to offer a few eggs or a lettuce or simply to exchange the time of day.

The *piazza* itself we hardly use ourselves. In summer it is given over to tourists who take photographs of one another sitting on the stone throne reputed to be Attila's. One wonders how he managed to mislay so heavy a piece of baggage in such a remote place as Torcello. Possibly he abandoned it deliberately because it was too cumbersome like those early pioneers of the American West who eventually had to leave treasured possessions like chests of drawers and grandfather's chair beside the Oregon trail. The *piazza*'s other frequenters are the glass sellers who have set up stalls on the grassy edge of the square and attempt to inveigle tourists to buy truly horrible examples of Murano glass. In winter a bitter cold wind sweeps between the basilica and the palace of the State Archives, now a part of the museum, whose *sottoportico* contains some charming fragments of capitals and Roman statuary found in the vicinity. The *piazza* is unwelcoming then and we charge across it, head down, into the shelter of the parish church of Santa Fosca. On Sundays we dress the *piazza* with a great billow-ing flag just like St. Mark's square in Venice or our neighbours in Burano. You did not think that we would allow ourselves to be out-done in ceremony or in the nice appreciation of symbols, did you?

During the summer months the path along the canal is black with tourists plodding along to visit the basilica and then back again to catch the excursion boat to Venice. This hardly disturbs our use of the path or the *vaporetto* stop as meeting places. Vene-tians usually walk fairly briskly and the people of Torcello are no exception, so that the locals tend to come in a bunch after the main tide of tourists has already passed. Nor do many of us make any attempt to get inside the shelter at the stop, preferring to stand outside in the shade of the trees talking to the large lady who is employed to marshal the tourists into some sort of order so that nobody gets pushed into the canal.

Very few tourists seem to have the time or inclination to explore the island. Some organized tours hardly give them time to reach the basilica let alone to look with any attention at the marvellous mosaics it contains. Even those who can afford to spend a whole day on the trip and venture to use the excellent and cheap *vaporetto* service seldom wander away from the main path. It is true that the little side paths mostly end abruptly at a canal or in someone's back-garden guarded by a resentful and noisy dog. But there are some entertaining walks. If one follows the path running along the far side of the basilica, hopping up and down banks and across little wastes covered with flowers, grape hyacinths, poppies, violets and sweet wild asparagus with the off-shore *barene* blue with sea-lavender in the autumn and speckled with white foraging seagulls, one will arrive at the outlying farmhouse of S. Antonio where one can look back on the apse of the basilica and the cheer-ful waterfront of Burano. Or one can turn off to the left from the main path along the canal before the first houses and strike across the common, or *bonifica* as they call it here, now uncared for and infested with tall reeds so that one expects to meet a rhinoceros as one emerges rather than the flat expanses of the lagoon. It is too shallow hereabouts for even a flat-bottomed boat to cross at high tide, and low tide reveals a tricky network of vein-like channels filled with rash little fish expecting to gasp their last if the water goes out any further. The lagoon on this side runs to the edge of the plain about Treviso and, on a clear day, the eye sweeps up to the Alps beyond.

Very occasionally one meets someone looking for a corner to sketch. They are usually ladies, no longer very young, clutching a pad, a paintbox and a folding stool. One finds them picnicking in the shade of a wild acacia, some with sandwiches and a flask of coffee propped up on a tussock of grass, others, more adven-turous, with some rolls and an *etto* of butter (now a little runny), some slices of ham and tomatoes and a bottle of wine imprudently bought in the local bar. They are often English and look so crest-fallen if one forgets oneself sufficiently to say 'Good morning' in their native tongue, thus demolishing an innocent pretence of having so melted into the scenery as to be indistinguishable from the locals.

Such bustle as exists in Torcello is on the water and is largely created by the *vaporetti*. Once the great Thomas Huxley told a friend, 'If I were not a man, I think I should like to be a tug.' If he had known the *vaporetti* which ply in the lagoon he might have changed his mind, particularly if he had known the last of the chugger-push *vaporetti*. There is now only one left and I am always delighted to be aboard her. She was built in 1939 and is considered the most reliable ship in the A.C.N.I.L. fleet. In general outline she is like all the other *vaporetti*, flat-bottomed, shallow-draught boats suitable in places where the low tide gives very little margin for manoeuvre. As the old boat comes up to the quay one recognizes her first from the persistent pounding of her green- and red-topped pistons. Then as she comes alongside with little short snorts punctuated by the staccato ringing of the electric bell relaying the bridge's commands to the engine-room, one sees the great housing amidships which conceals the engine. One goes aboard to a wonderful Conradian reek of oil and heat. On hot days the engineer opens the hatch and leans out mopping himself with a handkerchief in one hand and holding an oil rag in the other and by the end of a sweaty day there is not much to choose between them. One admiring glance past him into the engine-room, the brass gleaming, the paint shining and the running parts covered with a glaucous coat of oil, and one has made a friend for life. He is bursting to tell of his engine's characteristics and qualities and specifications and fuel consumption and horse-power and turns of speed and reliability and the tight corners it has got the ship out of and its extraordinary good temper and implicit obedience. He endows it with all the gentle sweetness of his mother and the robust manly qualities of his father. It is as obedient as his sons, as winsome as his daughters and as faithful as his wife. Chugging past Murano aboard her on a misty night, the glass furnaces projecting a more than usually satanic glow over the water, one feels that one has embarked with Jules Verne on some obscure and sinister journey. On a scorching summer day the trip from Venice seems a long and exotic voyage and one arrives at Torcello faintly grieved that grass-skirted girls have not turned out to greet one and that the island is not fringed with coconut trees swaying over a silver beach.

I once embarked for Torcello on this old *vaporetto* in a fog. There was some discussion amongst the crew as to whether they should start and, after telephoning for instructions, they decided to chance it. The A.C.N.I.L. (which is the municipal authority in charge of public transport inside the lagoon) has an enviable reputation for safety and punctuality and, now that all the boats are fitted with radar, they are seldom more than a minute or two late, even on a long winter run. But this was before the boats were equipped with radar and sometimes conditions were so bad that they could not complete the return journey and once or twice had even had to tie up to the stakes which mark the deep channels. We had no trouble, on this occasion, as far as Murano, but there-after the fog thickened and soon we were crawling from one marker to another. The navigable channels in the lagoon are marked by *bricole*, which are three tree trunks sunk into the clay bottom and secured together with chains. Visibility was soon reduced to half the length of the *vaporetto* and this meant that the captain, perched up on his bridge amidships, could not see beyond the bows. It was useless stationing a man in the bows because he could neither hear his shouts above the clicks and thumps of the engine nor see his signals because of the passenger cabin which takes up the whole of the fore-deck. So the captain stationed the deck crew on either side of the engine housing amidships and the dozen or so passengers on board joined them in their lookout duty. So we went forward dead slow, the calm water of the lagoon rippling past with hardly a sound. Then a shout would go up: '*Eccolo!*—There it is!' The captain would glance quickly at the pointing arm, strain to spot the marker and then call for a little more speed. As we passed the stake he reduced speed and we drifted tense and dead slow, until the next shout of '*Eccolo*'. Soon, being Italians, a spirit of competition combined with a kind of schoolboy delight at this little escapade, took over and the sharp-eyed vied with one another to be the first to shout. I, sharp-eyed but tongue-tied in the English manner, missed several markers that I could have claimed as mine. But there is nothing like competition (or was it national pride, or not bearing to be out-done?) for banishing inhibitions and soon I was yelling '*Eccolo*' as if I had been a lookout for A.C.N.I.L. for twenty years. Soon the short-sighted were banished into the cabin, including one of the crew who shrugged his shoulders at his own uselessness and went off into the warmth. The four or five remaining lookouts yelled and bounded up and down with excitement, the captain roared opprobrious comments on our parentage when we did not give him sufficient warning and the engineer, who had opened the hatch to join in the fun, kept up a running commentary from below decks. The whole scene was punctuated by the moans of the fog-horn and the constant ringing of the bridge's signals to the engine-room. We were all rather pleased with ourselves when we brought the ship safely into Burano after a delay of only two hours. I never got to Torcello that time, though I heard the great bell of the *campanile* ringing out through the fog like something out of the poem on the Inchcape rock. I stalked about the ghostly *calli* of Burano for hours and drank coffee in innumerable bars before the fog lifted enough for the *vaporetto* to start its return journey to Venice.

* * *

The old bed of the river Sile runs past my house and is the main channel to Treviso. It constantly carries boats of all kinds and sizes. There are the huge, lumbering barges, piled so high with merchandise that the helmsman, standing on a plank supported by two oil drums in order to see over his cargo, is on a level with my bedroom windows. When, after the disastrous floods in 1966, they were strengthening the *murazzi*, those great sea walls that protect Venice from the onslaught of the sea, enormous barges weighed down to the gunwales with blocks of dressed stone from the interior used to start coming past the house at first light. They were mostly propelled by sturdy little tugs whose robust chug chug chug is not the most restful thing early in the morning.

One of the most satisfactory things about the lagoon is that it is a place where people work and, being dependent on the tides and the seasons, work round the clock and round the year. The only exceptions to this dependence on the tides seem to be the *vaporetti* which, apart from a two hour pause in the early hours of the morning, run an hourly service to all parts of the lagoon; and the heavy barges which seem to keep industrial hours of work, though whether by agreement with trade unions or because of the times of opening the lock-gates into the river Sile, I have never discovered. Most of the craft plying in the Torcello end of the lagoon are small. Nearly everyone has a boat which they row in their traditional fashion, that is to say, standing facing the bows and with two long oars crossed at chest level. This is called rowing *alla valesana* or *alla Buranella*. Many are now powered by outboard motors. Of all the kinds of progress to which he has been exposed, the introduction of the outboard motor has made most difference to the life of the lagoon fisherman. It has greatly increased his range and consequently helped him to compete on more level terms with the new fish farms in the lagoon. It has greatly reduced those long hours of numbing cold getting to and from distant fishing grounds. His actual technique of fishing has changed little. He still scrapes up shellfish from the floor of the lagoon by thrusting a long-handled net along the bottom and levering it up across the boat to sort and throw back what he does not need. His method of getting small fish is the same by day or night, but more spectacular by night. Then one hears the rhythmic dip of an oar edging along the bank of the *barene* opposite the house with pauses while the fisherman pays out his net and pegs it firmly to the bottom with long stakes. When all is done he lights a bright acetylene lamp which he hangs on one of the stakes or, sometimes, well over the bows of his boat. He then rows about whacking the surface of the water with the blade of an oar to frighten the fish into the net. After a while he hauls in the net and tips the fish into the bottom of the boat and edges quietly along to another spot on the water's edge. Sometimes one can see five or six bright lights like will-o'-the-wisps creeping about the channels in the flat *barene* on a moonless night. By night and day, whenever the tide is right, they come to fish in this way within a yard or so of my windows. Out of the whole population of fishermen in the neighbourhood it is always the same two or three who fish here and I wonder whether by usage or by some immemorial tradition they have established a right to fish along the bank. And what would they do to me if I decided to do a proper job of fishing from my windows?

I was puzzled for some time when I first lived in the house by the faint but persistent sound of an engine ticking over. It took me some time to discover that the sound came from the throttled-down engine of a powerful launch lurking under the tamarisk which overhung the containing bank on one side of the house. I eventually discovered that the sound came from a Customs and Excise boat out for smugglers. It must have come to their ears that I had been making enquiries about the strange behaviour of the nocturnal boat, because I received a discreet message from them reassuring me and hoping that I did not mind their using this convenient hideout. Some nights later I had the satisfaction of hearing the launch's engines suddenly revved up as it shot out from under the bushes. I leapt out of bed and saw it arraigning a fishing boat which was creeping past silently under sail and without lights. There was a most tremendous to-do with searchlights and shouting and little figures leaping to and fro like silhouettes cut out of carbon paper. It ended with the fishing boat being ignominiously towed away. There seems to be a good deal of smuggling in the lagoon, no doubt supplied by ships plying in the Adriatic. The Customs and Excise people use helicopters as well as powerful motor launches and sometimes one sees one of them hovering over a fishing boat with the excisemen leaning out of the open door ordering the contents of all the baskets to be emptied out for his inspection and sometimes climbing down a rope ladder to inspect the inside of the lockers for himself.

There is also a good deal of poaching in the lagoon, particularly in the duck shooting season, and once we had a full scale naval battle to repel invaders of our fishing grounds. Although the lagoon is free to all there are traditional limits within which the fishing is reserved to a particular village or island. On this occasion twelve large fishing boats painted in a different livery from ours came beating up the wide channel from the sea. They were fishing boats from Chioggia at the other end of the lagoon and notorious poachers according to my *Buranelli* friends. Our boats from Burano put out to intercept them. They fight with oars on these occasions, sweeping and poking, and there were a number of broken heads and bloody noses and men pushed over-board before the police arrived in commandeered boats and arrested a number of fishermen on both sides for disorderly con-duct ('And piracy', yelled a number of my gleefully inaccurate friends). The men of Chioggia are said to trawl in a different manner from ours and are reputed to sweep our waters absolutely empty of fish whenever they come marauding. I suspect, however, that being normally deep-sea fishermen, the men of Chioggia are more efficient than our men. Then, local patriotisms run very high in the lagoon and the mere thought of a foreigner (let alone a hereditary enemy) in your home fishing grounds is simply not to be borne. However that may be, the bars of Burano were loud for days afterwards with excited, high-flown, raucous accounts of our brave battling boys and the repelling of enemy hordes.

The lagoon is the natural playground of the people of the islands. All through the summer the quays of Mazzorbo and the little muddy beaches of the *barene* opposite the Burano waterfront are alive with boys splashing about and diving headlong in one after another without a thought for pollution. The older boys and the young men practise rowing for the Burano regatta. My neigh-bour's sons, before they left the land, spent every spare minute urging a heavy, flat-bottomed farm boat up and down the canal by the house. By the time they graduated to a racing skiff painted a rather chi-chi shade of pale violet they were in such form as to take on any rivals from Burano or further afield. The ancient rivalry between the islands still shows in the eagerness with which the regattas are followed, though sublimated now from its earlier violence. In Burano the headquarters of the rowing club is still Romano's restaurant and there the history of the island's prowess may be followed in the championship flags displayed in their cases and the framed press cuttings of great victories. This is a famous restaurant for fish. Their fish risotto is particularly good and, while you are waiting for it to be prepared, it will be more worth your while to browse among the trophies of Burano's row, ing club than to inspect the indifferent paintings with which Romano's, like so many other Venetian eating places, chooses to adorn its walls.

On the day of the regatta coloured skiffs from all the rowing clubs of Venice and the lagoon congregate in Burano. Myriad boats of all kinds, from spruced, up vegetable barges and yachts from Venice to fishing boats turned pleasure craft, line the course. Henley cannot surpass this regatta in appreciation of rowing skill, in the earnestness of the competitors or in the gaiety of the scene. Nor can Henley match it in human warmth. The Burano regatta attracts a wide range of spectators. It has a sprinkling of gilded, be,yachted society, coolly bestowing its patronage. It also has almost everyone else who owns a boat, each one loaded beyond the safety limit with grandmothers installed in easy chairs amid, ships, wives under black umbrellas, aunts, uncles, cousins and fiancés, a brother,in,law's sister,in,law from Milan and a son on leave from the army, a bored young daughter languidly catching a roving eye in the boat alongside and innumerable children as skittish as kittens playing tig along the gunwales and never falling in, and large shopping bags crammed with food and drink. There they sit hour after hour, a little cup of strong black coffee in one hand and a bottle of *grappa* in the other, appraising style, rewarding courage and applauding success, particularly of their own colours. From time to time the *vaporetto* pounds past with a fearful list as all the tourists rush to the rail to record local colour in rather bleary prints with which they will later test the patience and courtesy of their acquaintances.

The lagoon also offers more solitary pleasures. The swing of the tide brings little out,board,powered craft put,putting past the house, conveying amateur fishermen to their favourite grounds. They are all kinds, lawyers, doctors, accountants, civil servants of all degrees, shopkeepers, hotel cooks, postmen from Mestre and

a *marinaio* from the *vaporetto* service. For it is to be observed that the
vaporetti are crewed by mariners. One buys one's ticket from a
sailor. The lagoon may be enclosed but it is still the sea and each
vaporetto trip is a sea voyage fraught with all the hazards which the
English, just like the Venetians, continue fondly to think the basis
on which their sturdiness of character is built.

From September till the spring, duck shooters join the procession past the house. They come past an hour or two before dawn
with their decoy ducks spread out on the decking of the bows.
The forked prop for the gun is already slotted into its holder. The
gun, still in its case to protect it from the spray, lies on a rug by the
bag containing some bread and wine and may be a sliced sausage
or a flask of coffee against the raw morning. They are mostly
alone. Some have a dog quivering with expectation and the cold.
Then all is quiet for a time. The first streaks of dawn bring the
duck flying in low in formation from the sea. They pass the house
with a swish of wings and a few moments later the pop, pop, pops
of the guns show that they have run into the hunters. Then any
duck passing the house wheel in dismay and talk to one another
and pick up their route again at a higher level, to be met also by
the inexorable pop, pop, pop in the far reaches of the lagoon.

The duck shooting in the lagoon is famous. The *valli* in the
back of the lagoon which lie in the path of the returning birds and
which provide the best sport are mostly private property and some
have belonged to the same families for many generations. In the
thirteenth century the shooting in the lagoon beyond Torcello
was reserved to the reigning Doge. Each year he was bound to
give a present of five wild ducks to the head of each patrician
family when they came to present their Christmas greetings. As
the patrician class increased the number of ducks required became
so large as to become a great burden on the Doge's private purse.
In 1521 nine thousand wild duck were required. To mitigate this
hardship and to ensure that the tradition continued, the Great
Council then authorized the Doge to have struck, every year at the
expense of the state, a silver medal to distribute in place of the
birds. These medals became known as *oselle* which is the Venetian

word for birds. An interesting collection of *oselle* is on display at the Correr museum in the Piazza. Apart from the private rights there is good shooting to be had for the price of a licence and membership of an association several of which have their own hides. Nearly every man on Burano has a gun and reckons to eat a few *selvaggina* of his own shooting in the course of the season.

The men who go duck shooting are on the whole proficient and merit the technical title of 'guns': not so those who wander about Torcello with a gun in their hands. Shooting off a gun seems as universal a sport in Italy as is fishing in France or watching football in England. At the beginning of each season the newspapers are filled with stories of fatal shootings, some of them so extravagant as to be hardly credible. There was recently the calamity which befell two eager sportsmen who went out together at the beginning of the season, and who quarrelled as to which had brought down a sparrow. They became angrier and angrier and finally one shot the other dead. We have our local Tartarins in Torcello who make life hideously dangerous at the start of the season by shooting any-thing that moves, even if it is on the other side of a locked gate. Giulindo was once shot at (and providentially missed) while inspecting the vines to see how the grapes were ripening by a man who thought he was a bird rustling in the leaves. Each year Giulindo and I have to bind up the shot wounds in the fruit trees in which some usually non-existent bird has been briefly perceived by a man licensed to shoot off his gun.

Apart from Sundays in September when sportsmen fire off their guns until nightfall, Torcello is a quiet place. It is so quiet that you can hear birds singing in far-off trees. One can no longer hear the nightingales. They have gone to less populous places. Torcello is like Assisi in that one should spend the night there to appreciate its quality. It changes its character when the last guided tour has left. Assisi seems to stretch, people bring their chairs out of doors and knit and quietly chat. Parts of the town that were shuttered and dead all day now open up and live. Voices come from dim passages and smells of cooking and children playing and the sound of a far off bell. Then Assisi belongs to the people who live in it. And so, too, in Torcello. The little bar fills up with local men, their wives come in to telephone. From as far away as the Devil's Bridge one can hear the water from the fountain splashing into the canal. Beyond the quiet and orderly bustle of Cipriani's, the glass stalls have been packed away for the night. You can hear the voice of the *arciprete* talking to his servant in his presbytery hidden behind Santa Fosca, or of the women sitting at their doors on the other side of the *piazza* straining their sight to put a few more stitches into their sewing before the light goes. Then Torcello belongs to those who live in it and if you want to taste its quality you must negotiate with Cipriani's for one of their rooms and go for a walk after dinner.

From the earliest times until the destitution of the Venetian Republic Torcello had been a great ecclesiastical centre. In 638 the Bishop of Altino transferred his see to the island where great numbers of his people had been established since they had fled from the waves of barbarians which had devastated the mainland cities in the fifth century. The great basilica which was then begun was as much a place of refuge as a church and the great stone shutters pivoting on leaden hinges may still be seen on a range of windows of the nave. By the thirteenth century there were seven great monasteries settled in the island with their churches, conventual buildings, gardens and orchards. To gain some idea of what Torcello looked like then one must go to the island of S. Francesco del Deserto. Here a tiny community of Franciscan brothers have maintained the order of life set down by St. Francis since the foundation of the monastery in 1220. The story goes that he was returning with his companion, the saintly da Rieti, in a Venetian ship when a raging tempest burst over their heads. As they approached the island the wind suddenly dropped and the waves sank into their usual calm. When St. Francis landed he was greeted with the song of birds and he built himself a hut of branches. He stuck his staff, which he had cut from a tree on the coast of Albania, into the ground and it sprouted into the first of the cypresses which, today, are one of the characteristics of the island. The monks still cultivate their garden, raise their bees, fish for their food and show their chapel and a cloister to tourists whom they also regale with the horrible story of the martyrdom of one of their number in China. One sees the brothers in Burano going about the monastery's business in twos and threes, or alone in a *vaporetto* going on more remote errands. They say in Torcello that if you see five of them in a boat it is going to rain. Theirs is a peaceful, remote and kindly society.

Monastic life was not always so. The Cistercian monastery of St. Thomas of the Burgundians was founded at the beginning of the thirteenth century 'in a remote corner of the island of Torcello' according to Flaminio Cornelio who wrote a history of the Church in Torcello. By 1390 it was a rich and proud foundation rejoicing in many endowments and becoming, it would seem, rather grasp, ing in its ways. That year saw the culmination of a tremendous, protracted quarrel which combined fighting off the claims of other monasteries to its estates and circumventing the pretensions of a strong Venetian aristocrat to filch the place of Abbot from its patrician holder to present to one of his nephews. This quarrel seems to have marked the peak of the monastery's prosperity. Thereafter it declined, as did the other monasteries with the change in the influence of the Church; as did Venice, with the decline of her trade after Vasco da Gama had discovered the sea route to India round the Cape of Good Hope; as did Torcello with the silting up of the surrounding lagoon which brought malaria and reduced the island's trade and population. Finally, after Napoleon had carried out his threat to be an Attila to the Venetian Republic and destroyed with it the relative independence from the Vatican hitherto enjoyed by the Venetian patriarchy, even the stones of the Church of S. Tommaso dei Borgognoni were shipped away to fill holes in Venice and its ornaments to adorn other churches there. The monastic buildings had disappeared over the years of decline. Nothing now remains except a small carved piece of marble which we seem to have mislaid, and the pottery shards which still litter the vineyard. The monks must have been great smashers of crockery.

CHAPTER VII

BUILDING STARTS

IN the beginning I used to think of the problems thrown up in building the house as the rungs of a ladder. One knew roughly how many rungs there were and that each led upwards to the next. When one reached the top rung the house was built. Eventually I learnt that the problems were more like the trees in a forest. There is always another tree behind or beside the tree in front of one and if one manages to make one's way out of the forest at all, it is only because one has met kind people who live on the spot and who lead one out through the winding paths to the plain where one can see where one is going. All kinds of problems unexpectedly arose relating to the nature of the soil, the type of building materials, the state of the ruinous cottage. There were unpleasant shocks like the ad valorem tax on the construction of the house which I only heard about two years after the house was finished and there were delightful surprises like being let off paying rates for thirty years because I had helped to 'improve' the lagoon. A generous and much appreciated regulation.

The plans finally approved by the Soprintendenza at the end of 1965 provided for a L-shaped house in the style of a lagoon farm-house. The entrance from the outside is modest and unassuming but once past the little lobby one is bounced into an Italian concep-tion of a house. It was soon borne in on me that the Italians incline to the grand (so do the French: consider all those mock Louis XV drawing-rooms) and also that they tend to be possessed by the idea of a thing rather than by the thing itself. A pale, honey-coloured marble floor (devilish slippery) leads to a beautifully simple stair-case with the treads and risers cut from single blocks of oak which give the same sensation of yielding solidity as one gets from patting the rump of a Suffolk punch or a well set-up Boxer dog. It needs a handsome Venetian cupboard and a Persian rug and then it will be splendid. Whether it will reflect the sort of person I am is another matter. The downstairs cloakroom shows how a concept governs reality. This space accommodates a place to wash, a lava-tory and the central heating furnace. The assumption which guided the planning was one of space and convenience. Conse-quently one enters through a door from the hall to the wash place. On the left is a door leading to the lavatory and to the right another door to the furnace. In an old Italian house one would take two or three steps to reach either inner door. This is the way one lived and so the notion continues though of course modern houses seldom have this amount of space to spare. The three doors just miss one another. One must shut the outer door before one can open the lavatory door or else leave it wide open into the hall. This I am sure was not the intention but is the alternative adopted by all except my most prudish acquaintances. To get oneself and a workman into the furnace room sets the same problems of ingenuity and the manipulation of apparently unrelated pieces as a Chinese box puzzle. The resulting complicated *chassé-croisé* sometimes irritates but more often amuses me since it arises from a quite unconscious pursuit of a national quirk on the part of these nice people. Side by side with this rather grand idea of how life should be lived goes a hard-headed practical appreciation of the way we do actually live now. The kitchen, for instance, opens conveniently on to the hall opposite the living room and next door to the loggia which are the two main eating places. It also, to ensure that we do not get too far from the notion of the grand life, has a full panoply of bells to summon me to the bathroom to go to my own assistance.

The huge main room or *soggiorno* occupies the rest of the ground floor. It looks on to the lagoon on one side and on to the loggia and a terrace on the other. Like all the rooms it is oriented to avoid the direct sun and has cunningly devised cross-draughts for cool-ness. It has eleven windows: three tall ones on to the lagoon, two on to the garden, two broad French windows on to the terrace and four in the *camino*. This traditional version of a lagoon farm-house kitchen is a small room built out from the main structure. It has a central stone hearth where the cooking used to be done and has seats built in round three walls. In fact it is a kind of large inglenook and is comfortable and warm on a bleak day. There is a charming old *camino* in Cipriani's restaurant. Mine is unusual in being so well provided with windows. Eleven windows present furnishing problems for there are hardly any walls against which to place furniture or to hang pictures. So far it has hardly any furniture in it and this rather bare state emphasises its pleasant proportions. Upstairs are my quarters composed of a large room with a little dressing room and bathroom, its own stone stairs and front door out to the garden and a top terrace which can be shut off from the rest of the house. Attached as I am to my friends and relations I like to be able to retire to my own fortress. There are three bedrooms giving on to a wide sitting room corridor with another bathroom at the end. The top terrace has views across the island to the basilica, the Alps and Burano. The bedroom windows look over the water towards the towers of Venice. My architect once described the house as *signorile*, which grand word means, if I understand it aright, not so much that it is lordly as that it is the kind of place that has a master bedroom. It is a beautiful house and it is very pleasant to live in.

Before all this could exist I had to find a builder. Two local firms were asked for estimates. Finally I came to an agreement with a small local builder who had worked on several projects in the lagoon with my architect. The specifications of quality insisted on by my architect left me dumbfounded. The window sills and outside door surrounds were to be of hand-hewn Istrian stone which has a marvellous, rough, dimpled texture. Downstairs the floors are marble or tiles. Upstairs, the floors and skirtings are of oak and the doors of walnut. The double windows, which keep the house warm in the raw winters and cool in the hot weather, are of golden Douglas pine. All this has cost about as much as a small house largely built of substitute materials would have cost in England.

But if the cost of the house was low by English standards it was still a lot of money to me. My architect took the line that if a thing was worth doing it was worth doing well, and that money spent on poor quality and skimped workmanship was money thrown away. I wholeheartedly agreed with him in theory but his comments did not help me with the practical problem of how one wed the cost of the best with a thin purse. I also had a new, and therefore unproductive, vineyard to keep going. And so my financial worries continued.

One day a team of men arrived to start building operations. The first thing to be done was to clear away the ruined cottage so that we could see the state of the foundations. Just then word came that the Soprintendenza was having second thoughts about the old structure and that an inspector was on his way to decide whether, after all, it would not be better to retain at least that part of the old house that faced on to the lagoon. It was most unfortunate that the walls in question happened to collapse into the lagoon just a few minutes before the arrival of the inspector. I was not in Venice the day this happened and when I next turned up on the site the whole team of men gathered round to tell me the story, with expressions of unparalleled innocence and in tones expecting a rebuke, of how the pile-driving machine imported to drive in the foundations had so shaken the old, crazy structure that the outer walls had fallen right into the lagoon. Look, signorina, there they are sticking out of the water. The only thing to do was to hurry on to the next problem (there is always a next problem) and pretend not to notice the subdued hilarity with which this regrettable *contretemps* was being recalled.

The pile-driver was certainly creating a disturbance crashing down every few seconds with a thump that echoed across the water. Its presence there was the result of telegrams reporting that the foundations of the old house had rotted away and had to be entirely replaced instead of merely strengthened as we had hoped. The lagoon is a constantly shifting world. One has only to observe the number of leaning *campanili* to realize that the floor is con-stantly moving, silting up here, nibbled away by the action of the water there, and carried away by the current from one place to be dumped by the slack of the tide in another. The replacement of foundations meant driving piles deep into the clay substratum of the lagoon. In my case we had to drive the piles six metres down. I have the impression that even if Torcello sinks below the water, like the neighbouring islands of Costanziaca (which disappeared at the end of the seventeenth century with all its monasteries and churches and thereafter served for many years as the depository of the bones of Venetians who had lain for the statutory period in the cemetery of S. Michele) and Ammiana (which had disappeared even earlier with several monasteries and eight churches), my house, arrogantly set upon its new six-metre concrete piles, will still mark the deep canal on the way to the lock-gate on the river Sile.

They say in Venice that a quarter of the cost of building or restoring a house goes on foundations. The ancient oak piles of the old house, almost fossilized with age, pressure and the action of the water had been eaten away by time and only a thread remained. They were certainly not robust enough to carry the weight of a large new house. The work has been very well done. The cost was enormous. With every thump of the pile-driver I saw my little heap of *lire* shiver and diminish.

Some weeks followed when nothing seemed to happen but holes in the ground. The pile-driver finally stopped thumping, was loaded on to a barge and taken away. A great hole was made and into it was inserted a prefabricated miniature sewage disposal unit of a kind which met all Venice's stringent sanitary regula-tions, trenches were dug to take the footing of walls, a great con-crete area was laid which let one see, for the first time, the size of the house, pipes for central heating ducts were laid criss-cross over it and stuck up incongruously where the inner walls were to lead them up to the next floor.

Then the house began to take shape. It was an exciting and alarming experience. I had no technical knowledge to help bridge the gap between the house I imagined I was building and what I saw actually going up. There was nothing in the apparently rickety, anaemic structure to prompt the imagination or to recall the charming drawings my architect had made. My heart sank as the house grew into different, all rebarbative, shapes: now like a pump house, now like one of those tall, raw brick structures which railway sidings are given to, now like a silo, improbably square. It could not get worse, I said to myself, but it did. The proportions were all wrong. The house remained obstinately tall and thin instead of being low and long and settling into the sur-rounding levels of the sandy flats and islands of the lagoon. And how could the proportions be put right when they were solidified in brick of a peculiarly repellent shade of pink? I had had from time to time from the very beginning of the exercise the notion that this was not something I was doing, but something which was being done to me. I went about Venice with almost a feeling of being caught, like a roving eye, dwindling at last into marriage with a heart lifting and sinking at the same time. For years, plan-ning houses had been an intellectual game; now, for the first time,

I was putting a plan into effect and look what had happened. I could not escape from the fact that it was I who had done this thing, who had allowed this horror to disfigure the lagoon. It was clear that it would be impossible, ever, to live in it. Since I was responsible ought I perhaps to pull it down: blow it up? What a way to dispose of one's patrimony! Then I thought of all the materials now assembled about me in dumps and of the beautiful houses I had seen composed of these same materials. Could I be wrong? Could I have misunderstood? I mournfully decided just to let things be for the moment. I would go away from the pink fright I had perpetrated on the banks of Torcello. I would go to Venice.

I have always liked to know how towns work, how they earn their living, set about their housekeeping, and how they adorn themselves, just as I like to know these things about people. Towns, like people, are a reflection of how they live. Biographies of both are normally unsatisfactory in this respect. What did Keats live on? We know he had difficulties with his guardian who no doubt cheated him, but this does not help us in finding out how he acquired enough money to pay his landlady. Wordsworth and Dorothy were so poor at Alfoxden that in the evenings they sat in the kitchen round one candle. But they had to have some money to live even in this modest style. Where did they get it? Money was not the most important thing in these men's lives and a biography filled with details of income and expenditure would in any case make dreary reading and miss the point. But in many cases it would be a legitimate aid to an understanding of them as minds at work to know what were the inroads on their time and nervous energy in getting enough to live on. In some cases we do know. Jane Austen was supported on the small means of her mother, the widow of a country rector and the household expenses were clearly helped out by her rich brother and this kind of life conditioned her work. We know that Scott paid off a huge debt by the earnings of his pen. It affected the quality of his work which is why it is interesting to know. Trollope tells us exactly what he earned and takes a solid kind of satisfaction in the accumulation of cash from his writing which is very typical of the man. These qualities are reflected in his novels. We know that Chatterton died in a garret impelled to suicide by despair and poverty. These things need no definition.

A town like Venice is a vivid reflection of the necessities of its life. It is a clean and healthy town, and has had to be healthy to survive. Fresh water is good and, until recently, has been plentiful, but the great press of tourists and of industry is now making excessive demands on the sources of supply. No sewage is allowed to flow into the canals. In the disastrous floods of 1966 one of the main horrors which struck Florence was the flow into the streets of the sewage from the *pozzi neri*, flushed out by the rush of water. This did not happen in Venice. The city has always been governed by stringent sanitary and health regulations which, in their essentials, are successfully enforced. The *calli* are usually swept and tidy though the increasing cost of wages and the difficulty of finding men to do the work are presenting the city council with problems. There are, there presumably always will be, corners where some charitable Venetian has put down a newspaper of fish bones for the communal cats and very unpleasant it is when the cats and the wind have spread them all over the *campo*. Household garbage is collected frequently though the plastic bags which, to save labour, have replaced the ingenious and characteristic tall bins which fitted on to little wheeled trucks for collection, have a depressing tendency to escape from the garbage boat and float off down the canal. There is also always some garbage caught under the bridges between tides, and men (though, I think, fewer than before) still go about in boats collecting it with large sieves fixed on the end of poles. It is difficult to prevent people from throwing things out of windows into a flowing canal below. It is done in a flash and even if there were an inspector on duty within sight it would be very difficult to identify from which window the orange peel was flung. I know how tempting it is to throw things out of the window. There is a certain pleasure in watching a crumpled up old letter go bobbing off on the tide to the Adriatic or, less satisfactorily, inland to Treviso. There is always the chance that one will see some silly fish take a leap at it. The ways of wasting time in the lagoon are pleasantly different from the opportunities afforded by other places. I have certainly not succeeded in preventing any of the people who work for me from casting things into the lagoon. There is a world of difference between oneself throwing out an old envelope and watching its fortune on the tide, and finding the discards of one's lunch (potato peelings, the tough leaves of let, tuces and the hulls of green peas) sailing impertinently past the sitting room on the incoming tide in the middle of the afternoon. One rushes out to expostulate. One must keep the lagoon tidy, one says. What about the stringent regulations, one says. Why was it not all put on the compost heap, one says. One is met by a charming smile. But the signorina knows that we lagoon people have always thrown our rubbish into the water and in any case, they say (suddenly remembering my liking for animals and other living things), it will feed the fish, *poverini*. As for the regulations, 'Eh! signorina!' The compost heap is regarded as either an amiable personal eccentricity of mine or as a mysterious fetish, to which I seem to attach great importance, imported from the misty north. Of course they put things on it from time to time (mostly tin cans and other uncompostable objects) to humour the signor, ina. But pure superstition it is, pure superstition, to think that a heap of rubbish can ever make things grow. What makes things grow is fertilizer out of a sack, everyone knows that.

Venice has throughout its history been a great seaport and a congested, over,populated place. Compared with other seaports round the Mediterranean it is not only clean, but healthy and orderly. For many years it was Europe's main link with the East and so it was often Venetian ships which imported devastating plagues into our Western countries. It was the first to suffer from these as the many thanksgiving memorials and churches bear witness. The Great Plague of London was said to have been started by two Venetian sailors who fell ill of it in the lodgings they had taken in St. Giles in the Fields.

Keeping the peace presents different problems in Venice from other places. There are, of course, the usual run of crimes. There are thefts for instance from open lower windows overlooking canals. Tourists, delighted by a romantic view, tend to find it less romantic when a long pole with a hook on the end, manipulated from a gently plashing row-boat, prowls about during the night catching up all the articles within reach of the windows. But the configuration of the town, the narrow *calli*, the frequent culs-de-sac, the sudden angles and the restrictive effect of the innumerable water-courses tend to prohibit certain types of crime and generally to work in favour of the forces of order. A bank robbery in the current smash-and-grab and fast-getaway manner is inconceivable in Venice. Similarly a rioting crowd can be very easily contained and controlled by a small force of police. This is no doubt why such student unrest as has occurred in Venice did so in the gardens surrounding the exhibition buildings of the Biennale. It is true that it was chiefly a protest against the Biennale itself and so it was logical to demonstrate outside it, but equally the space and un-restricted entrances and exits of the gardens made it a better and safer place for demonstrating and, by the same token, more diffi-cult for the forces of order to control. After all in 1310 it took only one old woman to frustrate the rebellion led by Bajamonte Tie-polo. As he was leading his men into the Piazza through one of the narrow entrances she is said to have dropped a flower-pot on the head of his standard bearer. The resulting confusion gave the authorities time to muster their men and put down the rising. You can still see her doing it from her window on the Merceria.

The sea is Venice's best ally providing protection, livelihood and health. The best cleansing agent, on which the city's health has always depended, is the scouring tide which twice a day sweeps the accumulated detritus of the town and the port out into the sea. The effectiveness of the tide depends on the swift action of the water and the replacement of the used-up water with clean, rejuvenated water from the sea. One of the major anxieties about the future of Venice is that the tide can no longer deal efficiently with the quantity of non-organic effluent which is being poured into it. This includes domestic detergents, heavy oil from domestic heating installations, industrial plants and ships. The outgoing tide cannot carry this new kind of pollution clear of the lagoon before the incoming tide washes it back again. The result is that these elements build up in the water, particularly in the little canals where the movement of the tide is, in any case, sluggish. So the oxygen content of the water, especially in the small canals, is becoming dangerously low. Apart from the effects on health and amenity, this has grave consequences on the ecological balance in the lagoon. Some kinds of fish are already deserting it.

Many Venetians also fear the effect of increased building and civil engineering work on the action of the tides in the lagoon. There is a whole corpus of knowledge about the tides and they fear that neglect of its principles and too great a confidence in modern technology may lead to disaster. In the days of the Serene Republic there were strict rules governing any building or exca-vating likely to affect the flow of the tide such as the filling in of water courses or the cutting of new canals. From time to time one comes across a short blind-alley canal called a *scomenzera*, from the Venetian word *scomenzar* meaning to begin. These canals are false beginnings where work, which was always undertaken piecemeal with intervals at each stage to check on the action of the tide, had to be suspended. It had been observed that a new ob-stacle placed in the path of the tide or a new inlet cut in the coast-line could have a catastrophic effect in deflecting the force of the tide on to the neighbouring quays or buildings. Under this action of the tide foundations have been eaten away, buildings known to topple and islands to disappear. These people say that the new deep canal for oil tankers being dredged west of Venice is being dug without regard for this old, hard-won knowledge and that it is already having a serious effect on the islands on that side of the lagoon which house the isolation and psychiatric hospitals. I do not know what the truth of this matter may be but I do know that the Venetians have a long folk memory for disasters and if their fears are groundless they should be assuaged.

Paradoxically, their dangerous ally the sea is also the Venetians' best protection. The tides are tricky and the navigable channels change quickly and treacherously. It was because of the dangers that the lagoon represented to the stranger that the first Venetians took refuge there against the barbarians just as in England in 1070 Hereward the Wake chose the intricate marshes of Ely as his base for harrying the enemy. In 809 Pepin, the son of Charlemagne, launched an attack on Venice from Malamocco, which he had recently conquered. It seems that once again it was an old lady who saved Venice. She pointed out to him an infallible way through the *barene* which was no way at all and there the Venetians hacked his army to pieces. I do not know in what ancient folio James Morris ascertained that she set him on his way with the well-known local cry of *Sempre diritto* but to any non-Venetian it is inconceivable that she should say anything else.

In Venice, transport also presents different problems from other cities. The *vaporetto* service runs like a leit-motiv through Venetian lives and its influence cannot be over-estimated in assisting the city into the twentieth century. It is cheap, frequent, punctual and, for water transport, quick. In contrast the transport of goods is slow, erratic and expensive. The difference in cost of exactly the same bottle of wine in Venice and Burano is 50 *lire* (3 new pence) made up chiefly of the cost of transport.

But it is the gondola which remains the queen of Venetian transport. This elegant town vehicle, this Rolls Royce amongst boats, must be framed in vistas of palaces and curving waterways to be fully appreciated. Like most things in Venice the gondola is not only what it seems to be—a convenient boat to get about in— but a vehicle of folklore, tradition and dreams. It has been painted black for centuries as a sign of mourning for a great plague, though Malibran the famous Spanish singer who stole the heart of Venice in 1835, and died so incongruously in a fall from her horse in Manchester a year later, went about in a scarlet gondola with golden trappings. Its decorations are sober in accordance with the sumptuary laws which in the sixteenth and seventeenth centuries sought to check the exuberance of the citizens' love of display, but some old boats still manage an ornately carved foredeck. To the tourist the gondola has acquired overtones of the *fiacre* of nine-teenth-century Paris. The disappearance of the little closed cabin which used to sit amidships has done little to allay a reputation similar to that of Madame Bovary's closed cab. The gondoliers still appear in tourist eyes to be tinged with the complicity of the cabby keeping the horse steadily trotting in the park through gazing crowds, alert with concupiscence and curiosity, imagining heaven knows what intimacies behind the lowered blinds. It must have been very difficult to conduct a love affair with a sense of anything but intrigue and turpitude in such a blaze of complicity —cabbies, ladies' maids, footmen and friends in the know. Perhaps it was the wish for greater privacy which led to the *hôtel garni*. It is, I suppose, part of our claim to live by a higher moral code than the French which leads us to describe a hotel of very temporary occupancy always in their language and to distort the phrase to give it a particular meaning which it does not carry in the original. This is all the more offensive to the French, who are a precise people, and who have a perfectly exact phrase to describe just this sort of hotel. Now amorous intrigue seems to have become mobile again and the sleazy hotel seems to be giving way to the motor car. As the town encroaches on the countryside and more trees are felled perhaps our societies for preserving them will be able to argue their usefulness for concealing cars so employed as a means of protecting them.

The disappearance of the *felze* or little cabins of black material stretched over a wooden frame with little doors and windows to protect one from the cold and wind and a picture of the patron saint to shield one from the dangers attendant on sea journeys, marked the end of the supremacy of the gondola as a means of transport, because it then became a fair-weather craft. It is a pity that this 'sociable hood' as Henry James calls it should have disappeared. It must have been very cosy. The *tende* or summer awning which protected the passengers from the sun also seems to have disappeared. This is strange because according to Horatio Brown it was invented for the comfort of foreign visitors to Venice and, as the gondola has become almost entirely a tourist plaything, one would have thought that it would continue to exist. The fact that they have gone is perhaps another indication of the changing nature of tourism in Venice. Today's tourist with less money to spend and less time to visit the city is less likely to take a gondola on his tours about the town than earlier visitors to the city. Most of them only step into a gondola to join the crowd of boats which process every evening up the Grand Canal clustered round the barge which holds the string band and the singers engaged for the nightly serenade. A lesser number hire a gondola to explore the most picturesque canals. A very few take a gondola in the spirit that they would take a cab in any other city, that is to say, simply as transport.

The Venetians themselves hardly ever take a gondola apart from the ancient *traghetti* services which ply across the Grand Canal from each of the gondola stations. The ancient guild of gondoliers was licensed to ply for hire by the Grand Council of the Serene Republic provided that each gondola station maintained a service across the Grand Canal. The *traghetti* are mostly still maintained. The gondoliers take it in turn to man this service. Customers mostly stand for the short trip and the cost is 50 *lire* which is the same as the *vaporetto* on a cross-Canal stage. From time immemorial the alighting passenger put his coin down on the forward gunwale: a trick, they say, he picked up from travelling with Charon. In the early days the guild of gondoliers was very powerful. Each section elected its head man or *gastaldo*, chose, licensed and disciplined its members, saw that they attended mass, looked after their own sick and aged, buried their dead, maintained the widows and put their orphans under the protection of one of their number. They organized feasts on their saints' days and suppers on holidays. They negotiated terms with the Venetian authorities and service in the fleet of the Serenissima. They even had banks to which all members contributed. In fact it was a well organized welfare state. With the passage of time the authority of the *gastaldo* and the influence of the guild waned though at the beginning of the century they still ran their own bank. Now all glory has departed and the over-riding influence in licensing and setting tariffs is the Venetian municipality.

The creation of the *vaporetto* service and, later, the licensing of motor-boats as taxis were severe blows to the gondoliers. They are a diminishing and not very contented fraternity. Their ancient skill, for so many centuries essential to the economic life of the city, is reduced to a minor adjunct, a fringe amenity, a snippet of picturesque folk-lore for summer visitors, like the creaking one-horse carriages which plod through the traffic of Rome. They naturally feel frustrated. In addition they hinder and are hindered by the increasing traffic of motor boats which create waves making the gondola difficult to control and splashing their clients with manifestly dirty water. They are now under-employed and have to seek other work in the winter. Even in summer they have to hang about too much, nursing their resentments. This, and the fact that they are not doing very well financially, shows in the maintenance of their boats, which are seldom bright with polished woodwork and steel with inviting brushed cushions and carpets. The gondolas are slower in the water because it now costs so much to take them every third week to the boatyard where they were careened and the bottom scraped free of clogging growths. The flower in the little vase on the foredeck is now too often plastic.

Like so much else in Venice the survival of this ancient trade depends on the Venetians themselves. I do not know a single Venetian who takes a gondola as a matter of course in the way that a Londoner, a Parisian or a New Yorker takes a taxi. If he needs to convey himself and his luggage to the station he will take a motor-boat. Even for family festivities such as weddings or christenings he is more likely to hire a motor-boat than a gon-dola. It is odd that the Venetians who are intensely—even obses-sively—patriotic about their city, who cherish its beauties and live its traditions in their everyday actions, should have turned their backs on the most famous and most characteristic symbol of Venice. Why should this be so? Is it because it is slow, out-moded and old-fashioned? Is it because they fear, as many of them will tell you, that the gondoliers are all Communists? Is it because they feel that the gondoliers make a good living out of the tourists or is it an expression of a general resentment on the part of the middle classes in Italy just now about the claims of the working people to better pay and living conditions? Whatever the reason the attitude of the Venetians towards the gondoliers seems too often to be one of offhand disregard. They consider them as a lot of layabouts sponging on the tourists. They do not seem to say these things about waiters or hotel porters or any of the other trades dependent on the tourist industry. I do not think the gon, doliers would sit idly by their boats or wander about the *calli* mournfully crying out 'Gondola! Gondola!' to drum up some trade, if they had anything to do. In plying for hire there are always doldrums but cabbies in other great cities of the world seem to be rather busy. In Paris, at least, they will tell you that they have to work eleven hours a day to make a living and that they do not suffer from many idle periods or *creux*. The gondoliers only ask to make a living. The fact that a group of them should have emigrated with their gondolas to the nearby seaside resort of Jesolo is an indication of their plight. It is a nonsense in practical, historical or aesthetic terms to use a gondola to provide trips for tourists in the open sea. The gondola is not built for the open sea. It is a lagoon and, more particularly, a town boat. But it seems to make sense, in economic terms, to the gondoliers who have tried it.

If the gondola is to survive in Venice—and it is essential to the tourist trade that it should—radical improvements will have to be made in two directions: to create conditions on the water to enable the gondoliers to manage their boats and to create reason, ably secure expectations of making a living. The large number of powered boats, *vaporetti*, barges, launches and pleasure boats of all kinds stir up waves in the Bacino (which is that part of the lagoon which lies between the Molo and S. Giorgio Maggiore) and in the Grand Canal which make it practically impossible to row and to control a fragile craft such as a gondola. How often one sees a laden gondola overtaken by the wash of a passing motor, boat crashing down on the waves with a force likely to break her back and the flat bottom flinging up spray all around. The gon, dolier can do nothing but steady her and wait for the wash to subside. Then he begins to row again until the next motor, boat creates another wash. His work is made extraordinarily heavy and difficult. His passengers find the jarring crash on the waves un-comfortable, progress lamentably slow and the wetting unpleasant. The life of the boat is also considerably shortened by this constant buffeting. It is not built to stand rough treatment. It is a lightly built craft designed to convey passengers quickly. Well handled in quiet waters it does just this. With two gondoliers it flies along.

The gondola is suffering, as we are beginning to discover that we are all suffering in different ways, from the technological society we have created. It does not help very much to say this because it is neither possible nor reasonable to reject the benefits of our new-found knowledge and to put the clock back to an earlier period, though one could find in Venice people who would like to forbid the use of powered boats in the city. It should, how-ever, be possible to find a compromise. There are already speed limits: they could perhaps be revised or more strictly enforced. There are already one-way canals: they could perhaps be extended and some restricted, at least during certain hours, to unpowered craft. The Grand Canal in particular could be closed between certain hours of the day to all except essential traffic (such as *vaporetti*, ambulances and police) and motored craft deflected to other routes. These are, after all, traffic problems found in every city of the world and there is no reason why Venice, simply because its streets are made of water, should not find solutions to these familiar difficulties. Some such organization of traffic would give the gondoliers the conditions in which they could manage their boats.

The other problem is to provide conditions in which they have a reasonable chance of making a living. This, it seems to me, is something which only the Venetians can solve. If all the Vene-tians who, when they visit Rome or Milan or other large towns in Italy or elsewhere, take taxis as a matter of course, were to under-take to hire a gondola three or four times a year throughout the year (because the out-of-season times are the most difficult for the gondolier) I suspect that the problem would be well on the way to solution. 'But why?' I can hear my Venetian friends say. 'We do not owe the gondoliers a living.' That is true, I would reply, but you do owe some practical contribution to the maintenance of the way of life of this city which you say you love so much, and the gondola is part of its life. The particular Venice you love, the Venice of tradition, will not survive unless you Venetians cause it to survive. In the first place because it is your city and you, the citizens, are the only people who can ensure that it is a living thing and not a series of beautiful but empty shells strung along a canal. In the second place, because the gondola is not so very much slower than other forms of water transport. Are you always, on every occasion, in such a hurry that you cannot afford the extra minutes to travel by gondola? From the Piazza to the station might take an extra twenty minutes, from your house to the Piazza, where you all constantly congregate, might take an extra ten minutes, and to the Fenice Theatre perhaps as long. Could you not afford forty minutes a year? Would there not be some satisfaction in arriving elegantly by gondola at the watersteps of the Fenice with your wife in her furs and your daughter looking her prettiest? Would there not be some satisfaction in thinking that you were actively maintaining the traditions of your city? And, in any case, you should remember that if enough of you hired gondolas the tariff could be lowered, while still providing a living for the gondoliers, and so you would have the satisfaction of helping your city at less cost to yourself. As the descendants of the merchant princes of old this should appeal to you.

Venice needs to restore her buildings to recapture something of her ancient glory. The Italian State is giving munificent help, so is the Municipality and so are many Venetians. One must hope that the governments of other countries will help and one knows that some of them are doing so already, and that also the many societies concerned with preservation will help and the many individuals who have been warmed by the beauty and the atmosphere of the city. But all this aid is for capital expenditure. It will not produce an annual revenue to help maintain the city's way of life. This responsibility must fall on the citizens of Venice. One practical way in which they could contribute is to give themselves the pleasure of hiring an occasional gondola to ensure that the craft of rowing and building them shall survive as part of the living traditions of the city. One may suppose too that, if the Venetians acquired the habit of travelling by gondola, something would speedily be done about the rough water in the Bacino and the Grand Canal. The Venetian *pater familias* does not care to be bumped about and his *signora* even less so. It might also lead to the resurrection of the *felze* whose use has only died out because the gondola is now hardly ever used in the winter. A small heater to keep the cabin warm is surely not beyond the wit of a Venetian to invent.

The tourist trade should also be called upon to help maintain a way of life out of which it makes so much money. Could not tour operators be encouraged to include in their prices an element for the hire of a gondola? Could not the hotels do more to encourage their clients to travel by gondola? The gondoliers would no doubt meet them on terms.

If measures are not taken soon the art of rowing a gondola will disappear. The average age of the gondoliers is rising. Young men go into trades with surer prospects. The art of rowing a gondola, of controlling it with precision and grace with its one oar in what Frederick Rolfe, Baron Corvo, in a characteristically precious phrase, calls 'the mode Venetian', can only be acquired by long practice.

The art of building this unique craft, which evolved over hundreds of years, will also disappear. As it is, the *squeri* or boat-yards where they are built are becoming fewer. There is a charming *squero* of unusually rural aspect just behind the church of S. Tro-vaso. On the other side of the narrow canal there is a walk with a most convenient parapet to lean over while one watches the work of repairing and building gondolas. Coryiat probably visited it when he lived in Venice.

Horatio Brown who lived for many years in Venice in the late nineteenth century and acquired an encyclopaedic knowledge of the history, institutions and skilled trades of the city has left an excellent detailed description of the building and furnishing of a gondola. I cannot rival his knowledge nor the clarity of his explanation and cannot do better than quote it:

'The first thing to be done in building a gondola is to choose the various woods of which the boat is to be made. The wood must be well seasoned, and without knots, if possible. These points are of even more importance in the structure of a gondola than in the case of other boats, for the planks of which the vessel is made are so thin that they are liable to warp, and the knots become loosened and start. When the wood has been chosen, the *squerariol* begins to lay down the gondola. The governing measurements, of greatest length and greatest width, are determined by four posts, placed at these main points. These measurements are permanent; and there-fore each gondola that is turned out of a *squero* resembles, in dimensions, its neighbours that come from the same workshop, though choice of wood, care in workmanship, and such infinitesi-mal variations as an eighth of an inch in depth or width will make all the difference to the speed and durability of the boat. The operation of building begins by setting up the stern and bow posts, the *asta da poppe* and the *asta da prova*, which are made of oak. The ribs, or *corbe*, of walnut, cherry, or elm, are then laid down—they are flat at the bottom, for the gondola is a flat-bottomed boat; and round the uppermost ends of the ribs, joining them all together, the binder, or *serchio*, of oak is fastened. At the points where the bow and stern deckings are to begin, two bands of walnut, rising in the middle, and called the *ponte fossine*, run across the boat, from one *serchio* to the other, and act as a counter support to the ribs, which might otherwise be pressed in by the strength of the binder. When this is finished, the hull of the gondola, as far as its strength and structural lines go, is complete. It remains to add the walls, or *nomboli*, of pine, and the *fondo*, or bottom, likewise of pine; the floor, or *pogiola*, rests upon the ribs, and protects the bottom, which is too delicate to bear treading upon without danger of starting. The decking of the bows used to be made of walnut-wood elaborately carved. It is more usual now to employ pine in large plain slabs called *fiubone*, divided into four compartments on each side by the *cantinele*, thin strips of carved or beaded wood. The little door which closes the decking in front, and makes the bows a safe storehouse for the gondolier's belongings—his oil, lamps, sand for polishing, coats, old hats, and dinner—is called the *portella*; and the two steps in the bows, by which one embarks or lands, are the *trastolini*. The *squerariol* has only to furnish the two *forcole*, or rowlocks, and the foot-rest, or *ponta piede*, the sloping piece of wood under the rower's hind foot, whence he launches himself forward to the stroke, and his part in the construction of the gondola is finished; for the oars, of beechwood, are bought elsewhere.

'It must be observed that the gondola is not built to lie perfectly flat upon the water; it is tilted slightly to one side, the side of the *forcola da poppe*, and is about an inch deeper in the water on that side than on the other. Moreover, if a straight line be drawn the whole length of the boat, from the *asta da poppe* to the *asta da prova*, it will divide the boat into two unequal parts. The side on which the hind rower stands is considerably larger than the other; and the *serchio*, or outer line, on the side of the *forcola da prova*, is longer and makes a wider curve than the *serchio* on the side of the *forcola da poppe*. This list is given to the gondola by cutting away the *corbe* on the side of the *forcola da poppe* so that the angle which they offer to the water, at the junction of sides and bottom, is more gentle than on the side where the hind rower stands. On that side, too, the gondola, as we have seen, presents a broader bottom to the water, and therefore more resistance; and thus the weight of the rower, which, if the boat lay quite flat on the water, would raise his *forcola* too high, destroy the balance of the gondola, and injure his control, is counteracted, and the boat is rendered perfectly obedient. . . .

'But there is a great deal to be done yet, before (the gondola) is complete in all respects. In the first place, there are the iron finishings for the bow and the stern. These are not made at the *squero*, but at some smith's shop, and must be thought about and bargained for separately. Each part of that singular beak, or rostrum, which ornaments the bows of a gondola, has its own name. The large hatchet-like head is called the *palamento*, or blade. Below the *palamento* come the six teeth, or *broche*, projecting out-wards; and between the *broche*, in each alternate space, are three *spine*, or steel pins, which help to fasten the *ferro* to the gondola. In a line with the topmost *broca*, but projecting towards the inside of the gondola, is another tooth called the *contra-broca*. Below the six teeth comes a long thin strip of steel which curves under the bows and clings close to the boat, and helps to keep the whole *ferro* in its place; this is called the *paletta*. These various parts complete the *ferro da prova*. The *ferro da poppe* is much simpler, being in fact a curve of plain steel, rounding and finishing the stern-post. A *ferro*, to be perfect, should have the edges of its *broche* in one straight line; but in these days of hurried workmanship a good *ferro* is not always to be found. They used to be made of hand-wrought iron, light and pliant, that would bend and not break if they came in contact with a bridge. Now, the new *ferri* are cast in moulds, and are heavy and brittle. A good gondolier will, very likely, possess an old *ferro*, which may have been an heirloom in his family for many years; for the *ferri*, if properly cared for and not allowed to rust, are indestructible, and will outlive many gondolas. . . .

'But, besides the hull and the *ferramento*, there is yet a third department to be considered before the gondola is fully equipped: the *felze*, *tenda*, *stramazzetti* or cushions, and *puzioli* or arm-rests, with their brass fittings of sea-horses, dolphins, harps, or columns. The *felze*, or little house in which the passenger sits, secure from rain or wind, is the most expensive item in the gondola. It is made of a strong wooden frame, thickly covered with woollen cloth, always black, and ornamented by tufts of silk or wool dotted along the roof. It has a door and two windows, and a little brass shrine for the picture of the patron saint. The door and all the inside is made of walnut-wood, stained black and richly carved, frequently with scenes from Tasso. The mountings of the *felze* are of brass: . . . The *tenda*, or summer awning, is a modern device, so modern that the more conservative among the Venetian families are slow to adopt it. It was introduced for the convenience of foreigners and is really extraneous to the gondola. Yet it must be admitted that the *tenda* adds considerably to the enjoyment of the boat in summer weather; and its pale creamy curtains, lined with blue, help to relieve that monotony of black, which is sometimes charged against the gondola. The carpet, the cushions, and arm-rests must be added, and then the gondola is complete. ... A young gondolier, just starting in life, is not likely to have a (sufficient) sum (to buy a gondola) by him; so the practice is to pay down a certain amount at once, both to the *squerariol* and to the *fornitore*, who supplies the carpets and cushions, and to guarantee the discharge of the remainder in monthly or quarterly instalments. When a gondola is new, it is left unpainted on the outside for the first year, as an intimation of its youth, and also as a sort of guarantee to any possible purchaser; the value of a gondola falls immediately after it is painted, for then it is difficult to ascertain the condition of the wood and the presence or absence of knots.'

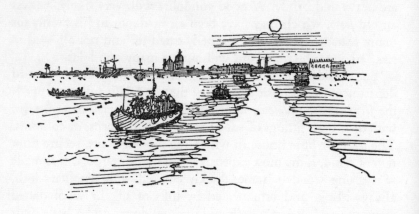

CHAPTER VIII

THE 'GANZEGA'

My next three-day visit to Venice was pure pleasure. Everything conspired to make it a success. The weather was magnificent, the lagoon sparkled, the Alps threw off the haze and rose up proudly on the horizon, the healthy little vines stood in neat rows spreading out their leaves in the sun. Everybody was good humoured and, most important of all, the house had acquired its proper proportions.

The tall gaunt structure which had upset me so much had given place to a long low house sitting solidly, but not obtrusively, in the landscape. There were two reasons for this sudden change. One was that the long wall enclosing the *loggia* and the top terrace had been built on to the main structure prolonging the roof in a gradual slope which entirely changed the profile of that façade. The other was that the builders' rubble that had littered the area had been banked up all round the house to make terraces so that visually the house started from the level of the terraces and not of the soil some three feet lower. The house was still a horrible raw pink, but I knew now that the stucco would deal with that.

It was a weekend of decision-making. This can be a harassing occupation, but not this weekend. Every matter requiring a decision had been well prepared and all the possible alternatives set out with the different materials and reasons readily to hand for consideration. We took all that Friday morning going round the house with the relevant *capi di lavoro* taking decisions in every department in turn. I had to decide chiefly on colours and textures.

We started with the outside. Nowadays most people in the lagoon colour-wash their houses. The advantage is that it is much cheaper than the traditional methods. The drawback is that it has to be repainted almost every year to remain neat and pleasing. The people of Burano, who take great pride in their gay houses and show an almost uncanny collective taste in choosing colours which, somehow, not only do not clash but, overall, harmonize, repaint their houses frequently. I had decided, given the hard winters and the salt-laden atmosphere, that it would be more advantageous in the long run to adopt the traditional method of mixing crushed burnt brick with cement to make a stucco. My architect and I arrived in Torcello to find that the new *loggia* wall had been newly plastered with five vertical stripes of different coloured stucco. A sack of crushed brick of the same colour stood at the foot of each strip. All the builder's men (unusually neat in dark suits and white shirts because later that day we were going to celebrate the *ganzega*) gathered around us vociferously offering their opinion on the five different colours which ranged from a muddy brown to an angry red. Finally we rejected them all, and the builder went off to the bar to telephone for more samples. The next day we arrived to find several more strips of plaster and we finally chose a crushed brick verging on crimson lake in colour. It seemed to me to have an unpleasing resemblance to the boot polish which, when I was a child, was known as ox blood, but I was told that this rather revolting shade would weather to the warm soft red which I had admired on the walls of the Cavallino town hall. And so it proved. By the next summer it had lost its blue-blooded look and then progressed through a fine sombre red, a rich tawny colour and dwindled to an elderly looking buff. Now one might think it was a hundred years old. I had been presented with a choice of textures as well as colours. I was astonished by the difference in effect between the various grades to which the brick had been crushed. The very fine textures gave a smooth, impoverished effect like cement. The very coarse gave an impression of botched workmanship, of being a *pis-aller*, and the next grade was pleasant but altogether too cottagey. We eventually chose a medium grade that gave a roughish surface which, when applied, gave a slightly pitted whorled effect. This rough-cast plaster is just right for this type of house. It avoids the sophisticated look of the smooth stucco appropriate to a town house at one extreme and an artificial straining after rusticity at the other. It was applied some five years ago and I still find myself inspecting it with satisfaction. This is surely a measure of its success. If one buys a picture or other object and eventually finds that one does not notice it any more, one can get rid of it: not so with the everyday dress of one's house. It has, in fact, proved to be worth all the trouble we put into it.

My architect and I then went up to the top terrace and thence to the roof, where we found the master mason waiting for us to discuss the three chimneys he had built. We had, earlier, held a long conference and taken a walk about the island inspecting other people's chimneys. Finally we sent him off with the architect's scribbled sketch on the back of an envelope as his only instructions. We were now to see the result of his handiwork. Two of the chimneys were to be of the tall oblong-shaped type topped off with a little tiled roof. These he had done perfectly. He had exactly caught the proportions of the stacks and the little roofs to the house. The third was to be one of the traditional square chimneys with the top part of the stack broader than the rest. He was the first to criticize its proportions. In fact he had already pulled it down twice and now wanted our advice as to what was wrong with it. This was a problem. It was very nearly right but he was quite correct in feeling that it would not do. We viewed it from all angles both from the terrace and from the ground and discussed all its elements in detail. He rebuilt it twice more before he—or we—were entirely satisfied. Now when he comes to the house, he is usually to be found on the top terrace in proud and satisfied contemplation of his handiwork. What a pleasure it is to work with skilled men who take such critical pride in their craft and how sad that the processes by which we mostly choose to live today give so few outlets for this creative kind of satisfaction!

The master painter was the next to claim our attention. Like the plasterer he had tried out various shades of green on the shutters of one of the windows on the landward side of the house. We stood down in the garden and he stood at the open window with a brush in his hand and two pots of paint in front of him on the window-sill. As we commented on his samples he added a little more green or a touch of black until we had found just the right shade of smoky grey-green, which has now faded to match the elderly olive tree. He then crossed off the rejected samples with two black strokes and painted all the shutters from the selected one. He came downstairs smiling with satisfaction at work well done. He too is an excellent workman.

We then went inside to inspect the floors. The entrance and hall had been paved with a polished, honey-coloured marble which beautifully set off the dark oak of the stairs and the faint mushroom tinge which the walls were eventually painted. No complaints here, work well done, compliments to all the men standing around and on to the living room. Here we had used twelve-inch-square red Florentine tiles, some of which had baked a little unevenly in the kiln giving a graduated smokey red in appearance. This offended the man who laid them. He suggested treating them with some preparation which would provide a uniform shade of red. We politely repressed him and instructed him instead to follow out the traditional method of swilling the room with linseed oil and shutting it up for four or five days. It is important that nobody walk over it during this period. The linseed oil sinks into the tiles and at this time they mark very easily and it is practically impossible to eliminate traces of footsteps. Thereafter we started polishing and, thanks to a friend who gave me a recipe for making up a wax compounded of virgin wax, paraffin and turpentine with an element of *carna ava* wax to provide a flint-hard surface, the floor soon acquired a magnificent dark glow. The *camino*, which is the size of a small sitting-room, had been given a floor of rough brick in keeping with its rural character and in the middle stood the stone fireplace under the hooded chimney.

As we were in the *camino* we started our inspection of the woodwork with the oak benches which line the three walls. Then we looked at the narrow skirtings and the double windows and doors and built-in cupboards throughout the house. I again distributed compliments to the carpenters and particularly on their skill in carrying out the architect's design for the stairs which combined the strength and solidity of a battleship with the elegance of a *sandalo*: I would say of a gondola, but this is a country staircase and should be compared with the country equivalent of the gondola. Upstairs the slim oak floorboards had also been stained in samples. We chose one that looked like old weathered oak and passed on to inspect the bathrooms.

Here, although it was not yet apparent, lay the germ of the only real calamity in building the house: the quality of the plumber's work. There had been some delay in delivering the baths, basins and lavatories and this had in turn held up the work of the plumber. And also the tiling of floors and walls which could only be done after the equipment had been installed. But the tiles were stacked on the terrace and it was with some excitement that I laid them out in blocks in the sunlight to examine the effect. My architect and I had spent time and care in choosing them. He had first of all shown me the bathroom tiles in his own house and also in Cipriani's charming and well equipped hotel installed in an old *palazzo* in the hill town of Asolo, which the Queen of Cyprus, Robert Browning and Freya Stark have, in their different ways, made famous. Then we went to a shop specializing in tiles in one of the back courts of Venice. Here the variety of tiles displayed to us was such that the glimpse of what I thought I wanted to do was lost in the profusion. The shop men threw themselves wholly into the problem of choice. Clearly choosing tiles comes into the same category as a *sistemazione* and consequently to be tackled with energy, imagination and a strict regard for practicality. 'Tiles,' they muttered, 'for two small bathrooms, predominantly blue.' They laid them out in blocks of four and soon the other customers were reduced to a kind of hopscotch to make their own purchases; and some lingered to admire and comment and, of course, to give advice. Eventually, I had made up my mind but I was still unsure as to whether I had made a good choice and I was glad to find the tiles so pleasant when laid out for inspection in the setting for which they had been chosen. The plumber, a chunky, muscular man with bright blue eyes and an engaging grin, had brought with him the measurements of all the equipment the architect and I had chosen together in Padua. We discussed, in detail, the positioning of each piece of equipment and he carefully marked the wall on each occasion. I went away reassured that the elegance and comfort of the bathrooms would go some way to compensate for their being so small. I had reckoned without my plumber. When I next saw the bathrooms they were fully equipped and the tiles had been laid on floors and walls. Only one piece of equipment was of the quality and design that I had ordered. The lavatories in particular were ugly, old-fashioned, badly designed and difficult to keep clean. Further, it had all been set in the walls like a dog's hind leg, as my grandmother used to say of things done without rhyme or reason. The towel rail was set immediately below the shower and the soap dish beyond the reach of any but the long arm of a gorilla. The man who laid the tiles had done his best with this confusion but necessarily he had had to slice tiles to accommodate this dotty arrangement of equipment. For months I regularly lost my temper whenever I went into the bathroom. There was nothing to be done about it except to strip the walls of equipment and start again. But at this point I had no money to do so, though one day I shall undoubtedly have to see to it. In the meantime it is a disincentive to having a bath. Later other, and more serious, deficiencies appeared. We found that the plumber, for some really unfathomable reason, had installed two hot-water cisterns, yet we never seemed to have enough hot water. It was only after months of puzzling about it that we discovered that he had only connected one cistern with the boiler. Still later the entire central heating plant had to be reorganized. The plumber had, it appeared, made no calculations of the space to be heated. Consequently, he had installed one-third fewer radiators than were required to heat the area adequately, but a boiler capable of heating at least twice the space. So however high one turned the furnace, the house was never warm. The fuel consumed to pro-duce this ludicrous result was immense and it finally proved cheaper and more satisfactory to replace the entire plant. But this distressing situation was not apparent on that sunny morning.

We were, indeed, all delighted with our morning's work and went to wash our hands in the water-butt before setting off to the *ganzega*. This is the equivalent of the English topping-out cere-mony: a traditional meal which brings together everyone who has had a hand in building and equipping the house, and its purpose is to wish the house prosperity and happiness. When we had washed we all climbed into the builder's old motor-boat to go to Maddalena's *trattoria* in the neighbouring island of Mazzorbo for the *ganzega*. The old boat had been scrubbed out and everyone but me sat on the gunwales on their handkerchiefs to protect the seats of their black Sunday suits. A magnificent black Danish-style leather armchair had been placed amidships for me and, sparing a thought for Enobarbus, I gave the word to cast off. The vibrations of the engine were so strong that everyone's outline was blurred and its clatter so great that we had to shout to our neighbours, but we maintained an infrangible dignity on the journey and on landing at the restaurant.

Maddalena's restaurant is well reputed in the lagoon for its cooking and is better known to the local people than to tourists. It used to be Sylvia Sprigge's headquarters and it was from there that she chiefly wrote her excellent book on the lagoon. She is remembered with affection and respect by the people who run it. It has now moved from the house that she knew to a new building further along the quay adjacent to the *vaporetto* landing stage at Mazzorbo. Twenty-five of us sat down in Maddalena's back room. The tables filled up from the ends towards the centre where the *capi di lavoro* sat round about me, and the talk too washed gradually up on either side like a tide, dispelling my anxiety lest the presence of a woman (and a foreigner at that) would make the men mind their manners rather than enjoy themselves. Maddalena's son ran to and fro with bottles of wine and soon the noise of our celebra-tion could be heard all over Mazzorbo. We had all Maddalena's specialities and she is regarded as one of the best cooks in the lagoon. We began with copious *antipasti*, ham, sliced sausage in great variety, sardines, anchovies, tomatoes, cucumber, radishes and sliced fennel. Then came great dishes of fried fish of many kinds (including those delicious *passarini* which look like dabs and taste so much better) and roast wild duck as the season had just begun. All these dishes succeeded one another at great speed and everyone helped themselves and their neighbours in whatever order of dishes they fancied. The gentry had their plates changed from time to time, but most of us ate everything out of the same plate. One of my neighbours absolutely refused to part with his fish plate on the grounds that duck ate fish too and so what was wrong with eating duck on a fishy plate. What indeed? The youngest of all the workmen, a fourteen-year-old stalwart who, in spite of his frail appearance, was capable and proud of doing a heavy day's labouring, collected *antipasti*, fish and duck all together and dis-posed of the lot with apparent satisfaction and many rolls of bread. We finished with plates piled high with cheese (*pecorino, gorgonzola* and *grana* amongst others) and dishes of fruit and a constant stream of coffee and *grappa*.

With the coffee and *grappa* we had speeches. My architect made a speech on his own behalf and then one on mine as my Venetian is not of this standard. Both were received with acclaim and I did not understand a word. Other speeches followed from each of the trades represented. Maddalena's son, flagging a little by now, kept the table replenished at a steady walking pace with wine, *grappa* and great jugs of coffee. As the afternoon wore on all enmities were forgotten and the plumber sang 'O sole mio' in a sweet tenor voice with his rival the electrician who was also a plumber and had competed for the contract. I was later to remember this and to reflect that I had not lost much by my choice for one was hardly better than the other at his trade. Then we sang all the current popular songs and all the old Venetian songs that anybody could remember and I was on the brink of 'Annie Laurie' myself when my architect and I had to go and catch the *vaporetto* to keep an appointment with the municipality. As we left the restaurant we found one of the masons bowed rather precariously over the table laden with empty wine bottles. I thought he was counting them: there were indeed an impressive number. But he was peering into them with one eye to make sure they were all quite empty. We put him in the charge of a friend who led him down the room to a better source of supply. All the other men came out with us and and stood on the bank and cheered as the *vaporetto* pulled away and we cheered and everybody else on the *vaporetto* cheered out of sheer pleasure at our happiness.

The next day was Sunday. I spent all day by myself, prowling about, purring, in my spanking new vineyard and my elegant new house.

CHAPTER IX

THE FIRST YEAR IN THE NEW VINEYARD

THE twelve months running from October 1965 were busy ones in the vineyard. As soon as the contractor had finished clearing and ploughing the land Fioretti and his men started work on the fence, which I had grudgingly to admit, now justified itself in providing some protection for the stocks of materials which the builder was beginning to accumulate preparatory to starting work on the house in the spring. Then Fioretti built the barn which, again, was useful at this stage for locking up the builders' more valuable pieces of equipment and the trees and vines which the nurseryman was beginning to deliver.

It was now that I took on Giulindo Berton. I found him through his brother who is a stone-mason at St. Mark's. Before coming to me he had worked for many years in the vineyards of the Armenian fathers, whose monastery of S. Lazzaro in the lagoon near the Lido was often visited by Byron. Giulindo is the youngest son of a family of ten whose father died when he was four years old. Life was a desperately hard struggle for the widow left with such a large, young family. She now looks a very old lady though she can hardly be beyond her sixties. She has that tremendous, forlorn dignity of the peasant matrons one often meets along the Mediterranean coast. It is a source of great pride to her that she kept her family together and launched them all in life. The only concession she made was to accept the offer of the Church to take the eldest boy off her hands. He is, as a consequence, the only one to have had any education beyond the age of twelve. Eventually he became a deacon and this connection with the Church is a great satisfaction to her.

Life is still very hard for the peasants. It is difficult for us, saturated as we are in a consumer society, to realize how hard it still is, how close to poverty are even their prosperous times, how few the facilities and comforts that we accept without thinking and how small and infrequent their pleasures. The Signora Clara, for instance, who looks after my house with such devotion, married very young soon after the Second World War and apart from her own annual children took pity on the orphaned children of a brother. They were thirteen in the family and so poor that they lived throughout one winter on *polenta* (which is a kind of stiff maize porridge) and dried figs. Naturally the children were often ill and this added to her difficulties. She once told me that she had never been dancing when she was young and has still never been to a cinema. She would not think of spending money on herself in this way. Now she watches television belonging to one of her sons. She commented to me lately on his liking to have a little meat for his dinner: 'Every night they expect the kind of food we used to have only at weddings. Times are better now.' Conditions have certainly improved but pleasures are still few and far between. I remember, when I once gave Giulindo's elder girl, then aged about eight years, a ride in my car, what a shock it was to discover that her intense, still excitement came from the fact that she had never been in a motor car before. I am glad to think that, on the return journey, she had sufficiently recovered to make faces at her astounded, footslogging friends.

Giulindo, now, is not unreasonably off compared with his fellows. He at least has enough to plan, and to put something aside, for the education of his boy, now aged four. Besides working for me for twenty days a month he works his own piece of land on the Cavallino *lido*. The holding has been in the family a long time but with sub-divisions between all the heirs on the death of each proprietor the area belonging to any one person is now minute. The usual large, barn-like house has suffered constant partition and is now inhabited by a whole tribe of cousins more or less closely related. Giulindo, as the descendant of the senior branch and as the only member of his immediate family to have remained on the land, has a little more room as he occupies their share as well as his own. There is electric light and an indoor water-tap supplied from the well which the proprietors clubbed together to pay for, but apart from this conditions are very primitive. He also has a little more land than the others as he works the share owned by his brothers and sisters. In all it amounts to something less than an acre divided into three small parcels each at some distance from one another. He certainly could not keep a family on this holding, which is why he works for me, but it provides a useful addition to his wages. He and his wife, Mistica, are skilful and hard-working and raise good crops of tomatoes, aubergines, peppers, peas and beans which they send to market together with peaches from about twenty little trees which are just beginning to produce. They live, almost entirely, off their land. Mistica is a very good cook and an excellent manager. They raise chickens and rabbits and every Christmas slaughter a pig, every

bit of which is used, fresh or cured, in one way or another, including a variety of sausages. Every summer they like me to take a mid-
day meal with them (with characteristic punctilio I am never
invited on the days he works for me in case I should feel that I
was being entertained in 'my' time). It is something I greatly
enjoy. Giulindo instals me at the head of the long table and sits
himself beside me. Mistica and the girls do not eat with us, though
the son Sergio is now considered old enough to have a place set
for him lower down the table: a sort of apprenticeship to the
reigning masculine society. The mother and sisters run to and fro
with plates and dishes, or lean on the back of a chair for a chat.
We usually begin with an excellent beef broth with little *pasta*
shells or stars swimming about in it. Then we eat the beef. Then
we have roast chicken or rabbit and, at some point at this time in
the meal, a huge dish of stuffed *zucchini* is left casually at my elbow
as Mistica knows how well I think she does them. We then have
cheese (usually that excellent *grana* which we know as *parmegiano*
and usually only use grated) and fruit and one of Mistica's
marvellous sponge cakes. The whole meal is accompanied by
home-made red wine and ends with coffee laced with *grappa*.
After this I am quite happy to lean back and listen to Giulindo
telling me all the gossip of the neighbourhood. Apart from the beef,
the cheese, the grapes to make the wine and the flour and sugar in
the cake, everything comes from their own land. I am aware that
such a splendid meal is to do me honour, and that apart from such
rare occasions, the family eat very frugally and, in winter, rather
drably. But thanks to their hard work, they eat better than most.

It is particularly sad in Giulindo's case that he did not have
more education. He is an intelligent, as well as a hard-working
and conscientious man and could have gone much further with
proper grounding. He has an orderly approach to his work and is
good with people and, given the training, would probably have
been capable of stimulating and controlling the productivity of a
whole workshop of men. In himself he demonstrates the whole
case for education as a factor in the economic development of a
country which is the purpose of the huge funds which are made

available by the United Nations to developing countries for building their education services. The United Nations is now so convinced that economic advance depends on most of a population having some education that during 1970 it made available from various sources over 150 million U.S. dollars to help create schools and colleges and to pay an international expatriate staff until the country concerned was able to train its own people to succeed them. The United Nations Development Programme not only sponsors universities and technical colleges whose influence on economic progress is direct and obvious but also assists secondary and sometimes even primary schooling. This is to encourage a rise in general educational standards to provide not only the skilled labour required but also a population capable of using the products of modern technology. Giulindo, in his own way, is convinced of all this too and recently told me, with great rejoicing, that the period of compulsory schooling in Italy has recently been extended by one year. This will make it easier for his son to get on. For himself, Giulindo does not at all repine. He is a most unselfish contented man.

He also has those endearing and maddening qualities which produce what I think of as the Figaro relationship between master and man. This depends on a recognition by each of his dependence on the other, combined with a good conceit of himself, and on the sense of equality and respect on both sides arising from this. This relationship was romanticized in many eighteenthcentury novels but it constituted something real, even if rare, which we in England seem to have lost long since, perhaps because of the greater separation of society in the urban conditions following the industrial revolution. If no one wishes to serve anyone else in a personal capacity any more in England is it not partly due to the industrialization of attitudes between master and servant in the nineteenth century, a carryover of the new factory attitudes into the home where servants became too often regarded as 'hands' banished to the basement and reduced to the status of skivvies? Compare Mr. Kipps' life as a shop assistant with that of almost any apprentice in an Elizabethan play or novel.

It seems to be a necessary part of the Figaro relationship that each should, on occasion, find the other quite maddening. If the one really cares about the interests of the other it is bound from time to time to take forms which the other resents. Figaro's master found him altogether too outspoken at times and too often right. One of Giulindo's most infuriating characteristics is to knock on the front door at half past six in the morning when he comes to work until I answer in order to make sure that I am all right. When I stick an angry head out of my window to protest he becomes very hurt.

'But, signorina, if I didn't see you I would not know that you were all right.'

'Of course, I'm all right. What did you expect me to be?'

'Well, something might have happened to you.'

If given strict orders to desist he will desist, but he does not really settle down happily to his work until he has heard me open all the shutters and put the house on a day-time basis. I, on the other hand, frequently madden him, when he has taken a private decision about the vineyard, for instance, to plant something in a particular place, by deciding that it shall be planted somewhere else. He argues a little and then he says:

'Haw! the signorina gives the orders here! I am only the employee! The signorina is the boss!' and, hunching his shoulders, he stumps off down between the vines.

I know I could not run the vineyard without Giulindo, and he knows it too. He knows that I know that he looks after it as though it were his own, insisting on quality, driving his sisters-in-law and his cousins, who occasionally work for him, as hard as he drives himself. When he knew I was in difficulties after the disastrous harvest of 1966, he offered to do without any pay in January and February of 1967 on the pretext that there was really nothing to do in the vineyard at that period. I hope that the mental mark-sheet which he no doubt keeps for me is not without similar entries to my credit.

The planting of the new vineyard was done by Giulindo, with occasional help from a cousin. It was a big job and took from the last days of October until about mid-November. Each of the four main sections of the vineyard was planted with a different wine grape, two red (*merlot* and *rabosa veronese*, local types that are doing very well) and two white (*verdizzio*, which is doing well, and *tokai*). The *tokai*, of course, produces a wine very dissimilar in quality to the Tokay made in Hungary. Given the difference in soil and climate this is quite natural but the quality we are producing is not as good as other *tokai* grown in the area. It has proved delicate and prone to disease, at least on my land. We have to give it twice the care we give to the other vines and this means spraying each individual vine carefully under the leaves eight times in the season. Giulindo is becoming very impatient with it and would like to uproot it all. It was probably a *tokai* grape that Federico, strictly against orders, decided to make into wine himself the previous year. The local inhabitants make their wine in great dark-green demijohns imprisoned in wickerwork cages. Perhaps Federico was not very good at making it. Perhaps the grapes were as unrewarding as the ones I have produced. Whatever the reason, the wine was quite disastrously nasty, thin, bitter and faintly tart. I suggested we pour it into the lagoon but one of my advisers said: 'No, we must find a buyer.' And he did. He sold it very cheap to a seminary. I cannot imagine saintliness or any fine religious quality coming from priests raised on such a wine. I hope I have not helped to breed a latter-day Savonarola to be a scourge to the faithful.

All these vines we planted in pairs along the line of supporting posts. The two cross paths were edged with thirty-six plum trees of various kinds. The vines and trees were so aligned that the tractor could run the whole length of the vineyard without check, which is a great time-saver. In the two small sections I planted thirty pear trees (mostly Williams) with artichokes between them and the rest of the 1,350 artichokes in between the vines. In one of the sections I put 1,650 asparagus plants. The rest of the land in the vineyard I reserved for cash-crops to be planted in the spring.

By the onset of winter the vineyard, planted according to the most up-to-date scientific principles, neat, spruce and without a weed, became an object of interest to all my neighbours who hung over the new fence on Sunday afternoons exchanging agricultural know-how.

The following winter was a very hard one in the lagoon, and the severe frost killed nearly all the artichokes which are Torcello's most precious agricultural asset. Since the artichoke does not begin to bear a commercial crop till its second year this was a calamity for peasant small-holders who depend on each year's harvest to see them through the year. As far as I was concerned this put me back a year only as my artichokes were in any case not yet bearing. So I started with a set-back as a reminder of the vicissitudes of the agricultural life. Georges Duhamel once said of civilization that it was '*une longue et patiente habitude*' and it struck me that this was true of agriculture too.

I had decided to grow cash-crops in an attempt to make the vineyard pay at least some of its way until the vines came into production. So, in the spring and advised by Giulindo, I planted 6,000 tomatoes, 4,000 aubergines, 1,000 peppers, 500 *zucchini*, peas, carrots, celery, ridge cucumbers, radishes, parsley, garlic and several kinds of salad.

That year the spring came late and consequently Torcello and the other lagoon islands lost their usual ten to twenty days' advance on the mainland which is due to the mildness of the winter owing to the surrounding sea. This advance is a precious advantage because growers in the islands can obtain the high prices which produce brings at the beginning of the season. When the good weather finally came the crops on the mainland were as forward as ours. There was much shaking of heads over the future.

Giulindo was as anxious as the others. He clearly wanted to do well in his first season with me. In many ways his is a pleasant job. He is, in effect, in charge of the vineyard and master of his work and time, apart from the monthly consultations when I come to Venice. He is in any case the sort of man who likes to do a good job. We had already had one reverse when we lost the artichokes in the bitter frost of the winter. He did not know me well enough to know whether or not I would blame him for this calamity or for a financial loss over the year's work which would probably result from our losing the advantage of the usual island mildness.

Perhaps it is because they exist on the borderline between sufficiency and hardship that the simple people who live around me show such kindness, such discretion in their kindness, such endurance and such resignation in adversity. The Italians are generally kind and discreet and perhaps the townspeople have these other characteristics as well. I do not know the urban working people (who are the most likely to need them) well enough to say. But the interesting thing to me was that, as I saw them practised in the lagoon, they were the necessary virtues of a pre-industrial society—and this within sight of the great oil complex of Marghera. Their kindness is a practical kindness and is an essential element in a society where a formal system of social security is still rudimentary by English standards (naturally enough, as Italy does not have the wealth to support such a costly, necessary service). In this sort of society one simply has to help others, because one needs help oneself in time of difficulty. I remember once when we had trouble with the tractor my neighbour, noticing that the engine had stopped, immediately abandoned his own ploughing to come to find out what had happened and whether he could help us to mend it. It was essential to do the ploughing then, while the weather held. He came to help in the knowledge that we would help him finish his ploughing if he could not finish by himself because he had spent valuable time helping us. And, of course, we did. Once when my mother, then aged eighty-four, was living alone in the house, she fell downstairs and hurt her head. She was discovered by the admirable sisters, Stella and Clara, who come every day, and between them look after the house as time and their family responsibilities permit. Stella stayed with my mother. Giulindo was sent to alert the doctor. Clara rushed off to get the boat. They put a wicker chair in the boat and settled my mother in it, swathed like a mummy in blankets. Then these two stout ladies, one in the poop and one in the prow, sculled her over to the doctor's house in Burano, giving a running commentary on the accident to all who asked what was going on all along the shore. This is the sort of attitude Richard Hoggart describes, in *The Uses of Literacy*, in a working⁄class society which needed to behave in this way for the same reasons. If you did not help one another nobody else would.

The feeling of loyalty to one's family (in the sense of the large, multi⁄generation family which is almost a society in itself) springs from the same need. You do not have to ask for the help of your family. You know it is there without question. A feeling of dis⁄cretion, a delicacy about seeming to interfere in other people's concerns, or having them prying into one's own, makes it prefer⁄able to keep obligations within the wide frontiers of the extended family. A different kind of discretion is inculcated into the English young, one which tends to work to other people's comfort rather than to one's own and I was interested to see the more frankly balanced attitude to discretion practised on Torcello.

However much you may be helped you have to endure your hardships yourself. The capacity to endure is consciously recog⁄nized as a virtue. On Torcello they certainly admire those who do not allow themselves to be defeated by circumstances, by illness, by adversity of any sort. This is the sort of toughness Giulindo's mother demonstrated when left with ten children under the age of twelve and her only means of livelihood knowing how to work in the fields. This endurance is usually accompanied by resignation, which is not fatalism (which allows one to opt out of making an effort to retrieve a situation) but a recognition that, try as you may, the odds are that you will not succeed.

This resignation arises partly from living close to the weather and the forces of nature. Fishermen have it in as large a measure as peasants. The elements to them are dangerous and unpredict⁄able and nature is not to be trusted. It is only townspeople or country people, who can choose their time to approach her, who can regard nature as benign. It is only these people who will permit themselves to be overtaken by a storm in the mountains and risk the lives of the professional guides (who would not take out their clients on such a day) to get them down again. It is only these people who will take a boat out in weather that keeps the fisherman at home and risk their lives, in the guise of lifeboat⁄ men, to bring them in again. The people who work on the land, or the sea or the hills, have a wary respect for the harshness of nature.

It is no surprise that the Romantics who discovered nature should have been both European and middle class, when not aristocratic. It is only in our moderate, temperate climate that nature could be regarded as a friend. One cannot conceive that anyone knowing the uncontrolled forces and fierce, veering moods of the elements in the tropics could regard nature as the friend of man or think that battling with typhoons and tidal waves and the long dreary soak of the rainy season as in any way ennobling to the character.

Nor can one imagine anyone who works out of doors, even in the temperate zone, a tiller of the soil, for instance, struggling to plough it when heavy with winter rains or reap it in the flying dust of a drought; or a shepherd, trudging over snowy hills wet to the skin and rheumatic for the rest of his days, viewing so harsh a taskmaster as nature with anything but a wary eye. They might lift up their eyes unto the hills, as the Psalmist says, but only because it was the quarter 'whence cometh my help'. The country gentleman riding round his fields in between showers, the parson visiting his parishioners on the pleasanter days of the week and other leisured persons walking off their dinners or journeying on foot to the neighbouring village (a pleasure we are too often deprived of in these days of congested roads) had time and leisure to admire and to philosophize over the surrounding scene. This is not to say that the working countryman has not produced a whole literature of his own, though I cannot help thinking that Robbie Burns is at best a sport and at worst something very near to a phony. He is perhaps most real in his reflection of the rumbustious side of country festivity. The main streams of traditional country poetry are work songs for almost every kind of activity, Christmas songs, where indoor snugness seems to play a large part, or the tumbling in the hay type of song which seems to reflect relief and joy that the long winter is ended, the brief, pleasant weather is at hand, so let us all enjoy it.

There are, there always have been, exceptions, working country people who had the sensitivity and inclination to look at the hills and trees and flowers and streams for their own sake. Perhaps, after all, Robbie Burns was one of them. I was once taken to see a vineyard at Sancerre by the *vigneron*'s mother-in-law, just because it was beautiful. She talked of the colours of the leaves and the beauty of the little pale green bunches of grapes and the scarlet of the poppies growing here and there. Then she suggested that we took the car to the top of a hill from which, she said, there was a very fine vista. When we got there she unerringly took us to the spot from which the view was best. Flowing down the hill were the fields and the vineyards of Amigny where she lived; over to the left the main vineyards of Sancerre swept majestically down the slopes; to the right fields rose rather steeply to end in a thick avenue of trees on the sky-line which hid the road. Atop the slope opposite stood the little hill town of Sancerre with its now vanished walls still marked out by the run of the houses. Of course there is sensibility in the countryside, even if it often expresses a measure of relief, but nobody but a townsman could have written *Meg Merrilees* and nobody but a member of the non-labouring middle classes could have thought that 'Peter Bell' in any way reflected what country life was all about.

Resignation to circumstances which they cannot change seems to mark his domestic animals as much as it marks the peasant himself. Donkeys, in particular, look resigned though whether this is because they have to put up with the same conditions as the peasant or because they are at the end of a long line of kicks I have never been able to determine. In the lagoon there is a donkey which is ferried about in a flat-bottomed boat to his various jobs. Perhaps it is a sense of the indignity of being carted about in a boat which adds to his sadness: he is a very lugubrious donkey. Working country dogs, too, often have a resigned look, perhaps because of their responsibilities as much as their condition. I know a sadly resigned and responsible shepherd dog. He works on the hills surrounding the village of Rouvres on the edge of Normandy about forty miles west of Paris. He is a large, black dog with a grizzled muzzle who is very happy to intercept you, escort you through the flock to his master who stands placidly leaning on a stick, watching his sheep graze and listening to pop music from a transistor radio slung round his neck. He is rather lonely and pleased to chat, breaking off now and again to send one or other of his dogs to round up sheep straggling into the nearby, unfenced, fields of crops. He has about three hundred sheep in his flock, some his own and some committed to his care by local farmers. This is the only itinerant shepherd that I know in this part of France and it is, I fear, a disappearing trade. In the Pyrenees of recent years I have met one or two wandering flocks. It surprises one less to find them surviving in that wild, brown, medieval landscape than within commuting distance of one of the world's great capitals. On Sundays at lunch-time the shepherd herds his flock on to some convenient place, instructs his head dog in his duties and goes home with the other dogs for his dinner and a siesta. The head dog is my black friend. Once I found him on a hillside bounded by paths running along the top and sides and a stream at the bottom. I sat on the hillside above him and watched him work. The dog stationed himself on the path above the sheep from which he could see everything on the slope. He sat at one end and then he trotted, very sedately, so as not to disturb the grazing sheep, to the other end, and sat, and then trotted back, and sat. At one point he saw some lambs down by the steam, running races and getting larky. Very quietly he trotted round the boun-dary path, not disturbing anybody, and interviewed those lambs and then he trotted back to his station, and sat. Soon after he had to trot down again because some fool sheep had got behind the village hall where he could not see what they were up to and emerged after a time with three truants strolling unconcernedly ahead of him, snatching bits of grass, *de la largeur de la langue*, as they went like La Fontaine's erring donkey. Then he trotted up again to his station on the top path and sat. It was then tea-time and as I got up to go I spoke to him. I was ignored. Not a twitch of an ear or an eyebrow acknowledged my presence. He had his instructions and three hundred sheep to keep contentedly grazing within given limits until his master returned. It was no time to talk to friends. I thought about this dog as I worked once in the garden in Torcello, supervised by a robin in his summer fustian. It is only the birds and the wild animals that are free. The labour‑ing countryman has learnt to accept with resignation the circum‑stances of his life and his never‑ending responsibilities for the crops and the stock in his charge. And he seems to have passed his attitude on to the animals he works with.

Peasants seem to find it difficult to adjust to changes in society because by the nature of their work and lives they are remote from the fountain‑heads of change which, in these days, are chiefly the government and the towns. I do not mean that they have any more difficulty in adjusting to ephemeral problems than their city contemporaries. To take an example, so small as to be almost absurd, the people of the lagoon have taken with good humour, and perhaps with more tolerance than the impatient townsman, the subterfuges to which they are reduced by the current shortage of small change. This has been parried by an insouciant, almost Heath Robinson, invention: you are simply given boiled sweets. The normal rate of exchange is one boiled sweet to five *lire*. In everyday commerce they are quite accepted, so much so that the tourist is seldom given an explanation. I have several times found the coffee trolley on Verona station afflicted by a shortage of change. The look of astonishment on the face of tourists at the sight of several coins and two boiled sweets on their out‑stretched palm adds a spice of cheerfulness to the bleak atmosphere of the morning after a night in the train. It seems to me, quite illogically, more surprising to get sweets in my change at Burano post office, though one would be the first to complain if government services were better supplied than the public. I do not know whether the exchange is valid in reverse. I have certainly never seen anybody proffer sweets as part of a *vaporetto* fare. I suspect they get eaten and so the final beneficiaries are presumably the sweet manufacturers.

It is the changes which affect the whole tenor of their lives which

are difficult for a peasant society to assimilate. This is partly due
to the extraordinary durability of their habits and ways of life and
thought. The persistence of time-honoured notions and their
influence on the day-to-day behaviour of even an intelligent man
like Giulindo never cease to astonish me. Crops picked during a
waxing moon are better than those picked when the moon is on
the wane and I have known him wait for an auspicious time. He is
always most particular when buying stakes for the vineyard (which
are young saplings some three inches in diameter) to make sure
that they have been cut from the western slopes of a hillside. They
are stronger and straighter and last longer, he tells me. He holds
these views so firmly that I sometimes find myself wondering
whether they are not perhaps true. But my normal scepticism
returns when he tells me that I should not sit by the open window
as draughts cause appendicitis.

These notions are an instance of the extraordinarily enduring
quality of peasant society. The peasants of Silesia are a case in
point. In contrast to many, they managed to remain owners of
their land after the devastating Thirty Years' War, whose horrors
are so vividly described by that finest of fictional reporters, Daniel
Defoe, in the *Memoirs of a Cavalier*. These same peasants, three
hundred years later, have managed to survive in better conditions
than most classes of society under the Communist regime in
Poland. Collectivization has not been pressed beyond endurance
in Poland, no doubt because of the protracted and outraged resistance it met in Soviet Russia. The peasants have consequently
retained a certain measure of responsibility, of choice and, even if
in a very minor way, some liberty in disposing of their produce.
Poland is a poor country and no doubt as its wealth increases
(which looks like taking a long time) agriculture will become
industrialized. This is probably a greater long-term danger to the
small measure of independence the peasants have preserved, than
doctrinaire principles of collectivization. In the meantime most
of the Silesian countryside shows the marks of a peasant economy.
The most obvious indication is that the fields are full of people
working. This is certainly the most striking feature to an English

eye and is in marked contrast to our own empty fields where one or two men with a machine reap and bind a whole harvest in a few hours. In the meantime the Polish drive to export its agricul‑ tural products constitutes some protection against draconian attempts to bring the peasants out of their pre‑industrial society.

The lagoon people have reached a confusing half‑way stage between a traditional and a new industrial society. There is now a large and growing population which commutes to work in the glassworks of Murano, to Venice and even as far as Mestre. The island of Burano and the little towns of the Cavallino *lido* provide a growing dormitory area and these are really people of an indus‑ trial society living in the country. The fishermen and the peasants who live alongside them are still at a pre‑industrial stage, governed by the tides, the seasons and the weather and to this extent must always remain task‑oriented, as the anthropologists say. But even they are having to compound with a time‑keeping economy: the times of markets and of *vaporetti* to get to the markets to sell their produce. As time goes by fewer and fewer remain independent. The fishermen take jobs in the fish farms in the lagoon. The peasants hire themselves out as agricultural labourers or, increas‑ ingly, leave the land. Employment brings them immediately into an economy governed by time for they are paid not by the task but by the hour. This new dimension to their lives is difficult to assimilate, particularly, but not solely, for the old, independent worker. There is a tremendous gap between the rhythms of work‑ ing for oneself and working for an employer which demands disciplines quite different from those imposed by the tides and the seasons. Federico, for one, never learnt them. But Giulindo, who is a much younger man, is acutely conscious of them. His watch is a treasured possession and he always wears it unless he is doing something which is likely to injure it, such as driving a vibrating tractor. He uses the *vaporetto* to get to and from his work and, as he cultivates his own land too, it is important to him not to miss the *vaporetto* that will enable him to put in a little work at home before dusk. On the other hand he never skimps his work for me and, being the conscientious man he is, the needs of my crops come before the needs of his. This cannot always be the case. The security and escape from responsibilities which employment brings must often be accepted without recognizing that it is accompanied by demands for punctuality and a more continuous effort throughout the hours paid for. In ours, the oldest industrial society, it seems to have taken generations to inculcate these new, necessary characteristics of a factory life. It will presumably take at least as long in Italy, especially in those areas of employment, like agriculture and fishing, where the external aids to the new discipline, such as clocks to be punched, are absent.

So, because they have arrived at a difficult stage, it is no surprise if acceptance of these new time measurements in their work and lives is somewhat haphazard among the lagoon people. It may be a necessary evil that they must accept to get a little richer or at least not to get any poorer, but I cannot think that in the long run an approximation of all our lives to the rules that make a factory profitable will make our society any pleasanter to live in.

Yet this is what we are offering to our own people as much as to others: a costeffective society. What will happen to our ancient concept of personal liberty in a society where men skilled in the usury of time, as Wordsworth said, whether technologists or politicians, will use the supposedly inexorable discipline required to maintain a highly productive society to exploit us all? It seems to me that much of the protest of young people today is really directed against this rather than a consumer society which is merely the outward and visible sign of the inadequacies of the other. It is perhaps worse that the advanced countries should be offering it to the developing countries with all the weight of authority of their technological knowledge and with such persuasiveness and persistence as almost to be imposing it on them. The trouble is that the aid always reflects the attitude of the donors. Sometimes it is used consciously to infiltrate an ideology or to support a policy. Sometimes it is only unconsciously that the giver exports his attitudes but he does so all the same since it is those attitudes which have governed his choice of projects to support. Aid is becoming suspect to the developing countries for these reasons. But a much more insidious danger for them lies in the fact that behind the ascertainable motives for giving aid there are concepts so obvious to the donor that he has never stopped to consider their effects on his society, let alone on a different one. So we confidently invite the developing countries on to a slope which we are beginning to find so slippery ourselves.

It is bad enough that the changes we are embarked on should create such strains in a society we have moulded, over the centuries, to fit our own particular needs. It is at least something that we in the West, at any rate, have freely accepted, though without understanding (and possibly living to regret) the constraints which seem to be implicit in a society which reduces people to productive little gnomes. It is surely worse to impose on other people what we are doing to ourselves because, in most developing countries, their forms of society which also reflect the different identity of peoples will not be changed but destroyed, in the process of constructing this new cost-effective society. The arrogance of those who possess the secrets of technology has already created a state of anxiety in the minds of those who do not. They feel inferior, diminished in their humanity and even, as the society on which they depend for their sense of identity becomes increasingly disconsidered, becomes a non-society, so they themselves become a non-people, without an identity. The first step has already been completed. Most developing countries do feel inferior to industrialized societies. So they are, if the value of a society is reckoned by its manufactured products. But are motor cars and refrigerators worth anything unless a society also has a Shakespeare, a Michelangelo and a Plato? And are not the Shakespeares, Michelangelos and Platos of what we are pleased to call the developing countries as important in their societies as these three are in ours?

All this creates very complex situations for the people concerned. Some, seizing on the arguments proffered by Communist or indigenous extremist propaganda, find in the greed of their past colonial masters the reason for their present poverty and technical backwardness. Others, more sensitive and so perhaps more shattered by the implications of the disregard in which their society is held by 'progressive' nations, seek their identity in the culture which bred them. This explains the force and persistence with which they insist on the breadth and subtlety of their cultures and on the need for other countries to respect and learn about them. This was one of the main forces behind UNESCO'S ten-year project to create better understanding between Western and Eastern cultures. The Africans are in a particularly difficult situation since they do not have a written record of their past. If they are to know the kind of people they are, they must record the oral tradition before the old men disappear, who store it up like the Druids that Caesar described. If they do not succeed in recording it they will have to create their national identities from tomorrow. It is a terrible thing not to be supported by a tradition and by the sympathetic warmth of a fellow feeling, even if its main use to one is to react against it. I once heard a wise old African say: '*Lorsqu'un vieil homme meurt, c'est toute une bibliothèque qui disparaît*', which loses something of its quality in the rhythms of another language ('whenever an old man dies, a whole library disappears'). Of all the forms of aid offered to Africa the most important is help to record their oral lore for it is the safeguarding of the record of themselves: their identity.

The lagoon, like all things Venetian, is supremely confident of its identity and this confidence no doubt helped it to forget its anxieties in the pure joy of the sparkling sun and the trees dressed in blossom of the late spring. I went there as often as I could and those weekends (which are the rags of time) spent planning the new garden in the midst of the builder's rubble, are amongst those that stick most in my memory.

By June, the vineyard was a well-ordered paradise, a garden fit for the eye of Andrew Marvell with the sombre greens and purples of the aubergines like heavy dragoons amongst the bright scarlet and yellow light cavalry of the tomatoes and peppers. The summer soared on, with blue skies and occasional light rain at night. Giulindo sweated, picking, packing, despatching, propping up

heavy-laden plants. Then the markets started refusing produce—
never had there been such a glut. Under the incessant chatter of
the tourists a kind of hush fell on the lagoon. Nobody was hungry
yet, but what of the winter?

CHAPTER X

THE GREAT FLOOD AND THE
FUTURE OF VENICE

ON November 4, 1966 there were tremendous cloudbursts over
the Appenines and the Alps behind Venice. The water of the
River Arno swept through Florence with such force as to wreck
houses, museums and libraries, and devastated the city with its
violence. In Venice the effects were, at first sight, less dramatic but
laid bare the sapping menace of its watchful and neglected enemy
the sea.

In Venice the phenomenon of high water or *acqua alta* is not so much a high tide as a tide which is exceptionally or inconveniently high for this city seated upon the waters. The *acqua alta* is governed by three elements: the moon, the *scirocco* or hot south wind, and the *secca* or oscillation of the Adriatic sea. The height of the tide in the Adriatic, as in all seas, is influenced by the state of the moon and is at its highest at the periods of the new and the full moon. The effect of the *scirocco*, blowing straight up the Adriatic, is to bottle up the tide inside the lagoon acting as a valve which allows the tide to run in but prevents it from flowing out through the narrow channels which link the lagoon to the sea. On November 4 it had been blowing for three days. Indeed all that summer the *scirocco* had blown intermittently, enveloping the lagoon in a clammy haze. It had never been so persistent in living memory. At the same time the *secca*, the strange phenomenon which is met with in some lakes (the Lake of Geneva is one) as well as in the Adriatic, and which consists of a periodic oscillation or upward and downward swing of water as on a see-saw, had tilted up at the lagoon end of the sea with the effect of raising the level of water in the northern part of the Adriatic and the lagoon. In addition the storm water pouring off the Alps broke the banks of the rivers Sile and Tagliamento, in particular, and flooded the low-lying hinterland of Venice. This flood water was not a direct factor in the disaster in the lagoon as the rivers which used to empty into it had long since been deflected into separate channels. But it was a factor in isolating the city and in devastating the farm lands in the area between the lagoon and the high Alps, and provoked an economic disaster for the whole province of the Veneto.

In Venice, November 3 was a day of blustering wind. Normally the *scirocco* is a gentle clammy breeze, enervated by its long flight over the Sahara and the warm waters of the Mediterranean, but on that day and the next it was to reach up to 100 kilometres an hour in boisterous gusts. The water began to rise at eleven o'clock at night. By midnight the Piazza and the city centre generally were awash. By one o'clock on the morning of Friday the 4th, the water seemed to be receding, three hours later it seemed to have reached slack water, though it was still very much higher than normal for that period of the tide. Then it began to rise again. By seven o'clock in the morning of November 4 it had reached its maximum height of 1.89 metres with the *scirocco* blow, ing behind it without intermission. This is the highest recorded flood in Venice. The last three high water marks were 1.53 metres on January 15, 1867, 1.47 metres on April 16, 1936 and 1.51 metres on November 12, 1951. The flood maintained its maximum height for twelve hours. At seven o'clock in the even, ing of the 4th the wind changed and the flood water moved steadily out of the lagoon through the three narrow openings into the Adriatic sea. The rushing tide, the wind and the torrential cloud, bursts left a trail of frightful ruin throughout the province of the Veneto.

The Venetians know about floods and took energetic and expert action. There was no electricity, no gas, no drinking water, no bread for forty-eight hours, no telephone to summon help. They were on their own and they reacted with all the vigour and capacity which once made their city great. Neighbours and friends, as well as the police, *carabinieri* and firemen, organized themselves into groups first to rescue people, then to distribute food to families caught in upper floors and finally to clean up the town. Their first duty was rescue, particularly of the old and the poor because it is the old and the poor who are reduced to living in the damp insalubrious dwellings of Venice's ground floors. Here the weight of the water had crushed in doors and windows and driven the people trapped in them to climb wet and shivering on to the tops of wardrobes and dressers calling for help. The high water made rescue difficult and ambulances could not always negotiate the low bridges. Many old people finished their journey to hospital hoisted up on a young man's shoulder. Then the volunteers improvised temporary quarters for those washed out, found offers of hospitality for others and conveyed whole families to the flats of more fortunate relatives whose houses had not been damaged. They organized the distribution of food, hastily collected from those who had some to spare, for those marooned in high flats, whistling in the *calli* until the housewives let down their baskets on a string to receive the meagre contribution that was offered.

Then they turned to help the merchants, many of whom had lost all their stock. Shops are necessarily at street level and in Venice they are also usually small and crowded in every corner because of the lack of cellar space. Bands of young men made chains to convey what could be saved to temporary storage above the flood level. There were incongruous sights. Boxes of fruit, cases of goods, bundles of irreplaceable personal belongings were to be seen sailing off down the Grand Canal. The course of the outgoing tide could be traced along the *calli* in the long trails in the water of blue, red and green paint which started in the flooded store of a colour merchant near Rialto. The merchants also suffered from their shop windows being smashed partly by the weight of water but also by boats, torn adrift by the wind, which bumped and crashed along the *calli* until captured and led back to their moorings. Over seven thousand shops, business premises and arti-sans' workshops had been damaged in the city centre. The damage to houses and to miscellaneous property was also enormous. Many boats were wrecked with consequential hardship to their owners whose livelihood they were. Amongst them were many gondolas totally destroyed by the wind and waves.

The police and fire brigade had been constantly on duty since the water first rose on the Thursday night. There was, luckily, only one serious outbreak of fire due to a short circuit. There was no looting or the sort of disorder that one might expect in such a situation, though there was, of course, some cashing in on the possibilities offered by the occasion. Thus according to the *Gazzettino*, the Venetian newspaper which managed the extra-ordinary feat, in the circumstances, of appearing as usual except on November 5, the day following the flood, a private boat service was organized to carry people to their homes along the flooded *calli* at somewhat exorbitant rates. Others profited to a more modest degree by offering to carry people piggy-back to their houses.

The villages in the islands of the lagoon and along the *lidi* which divide it from the Adriatic suffered the same kind of conditions as the city people except that the flood came upon them more swiftly and harshly because they were less protected from the winds and waves; which latter, at least, had to some extent expended their force by the time they reached as far into the lagoon as Venice proper. At S. Pietro in Volta situated on the containing *lido* not far from Alberoni the sea wall suddenly collapsed under the raging sea and the villagers had no time to do more than take to their roofs and to some of the larger boats which could live in the stormy waters of the lagoon. Three thousand of them were forced to take refuge in the local barracks, hospitals and schools. S. Erasmo, the low-lying market garden island which lies immediately opposite the lagoon entrance, was suddenly and entirely overwhelmed by the rush of huge waves forced along by the raging wind. From his upper floor in his house near Treporti on the other side of the lagoon entrance, Giulindo could see no land whatever as the sea had drowned the whole breadth of the *lido* which, at that point, is about two miles wide. Many peasant families round him had taken their cows, calves, pigs, rabbits and chickens as well as their families to the upper floors of their houses to keep them safe from the flood.

Outside the lagoon two areas of the Veneto were particularly hard hit. They were the high valleys of the Dolomite Alps which bore the brunt of the cloudbursts and the torrential rains, and the flat farming land between Treviso and the coast which suffered the simultaneous onslaught of the sea before them and the floods arising from the storms behind. In both rescue operations, measures to safeguard health and the distribution of food were made particularly difficult by the collapse of communications. Roads were blocked, bridges were smashed, railway lines were torn up and telephone posts uprooted by the force of the water. In both areas the army and the police worked day and night clearing dangerous buildings, improvising bridges, evacuating entire villages, identifying the dead and taking the injured to hospital.

The resort towns along the coast were badly buffeted. In Jesolo sea water driven on by the high wind swept two miles inland and the townspeople had to take refuge in the top floors of the hotels. Further along the coast at Caorle the sea breached the defending wall in five places and rushed through the streets in a six-foot wave like the bore on a great river. Inland, the river Piave broke its banks at S. Donà and further inland the river Tagliamento broke its banks at about the same time. The rush of these released waters was so strong that it bowled houses over and its speed so great that it overtook people running out of its path. The roll of dead and injured began to mount. Over 120 square miles of fertile farm land were under water. Ninety per cent of the cattle in this area were drowned. From the road running along the 'levée' closing the northern side of the lagoon the plain of Treviso was flooded as far as the eye could see. At least it was sweet water which flooded this land and in spite of the tremendous damage done the soil itself was not poisoned by salt water as was the case elsewhere, particularly in the lagoon.

The cloudburst had cut communications with the mountain valleys and it was not until November 6 that news began to come through of the appalling calamity which had struck down the population there. Here it was the ferocity of the storm and the constant downpour of rain which did the damage. Some smiling Alpine valleys such as Fiera di Primiero had in the valley bottoms gouged out by the force of the torrent which left a waste of boulders and mud that will take many years to rehabilitate. In other places the rain seeping through the hillsides loosened huge landslides which blocked the roads for stretches of half a mile at a time. An avalanche of water, earth and boulders swept away houses at Mezzano and lay around the village in a layer ten feet thick. Here the villagers pooled their remaining maize flour and cooked huge collective *polentas* in the little available safe water. The hamlet of Vallanza was entirely covered by debris from another avalanche of mud. The only way to one hamlet was across an improvised bridge, through deep mud for half a mile and then a five-hour walk over the hills. Three valleys in the Belluno area were entirely cut off.

A brigade of Alpine troops and army helicopters was sent to the rescue. They evacuated casualties, surveyed the damage and re-established communications. They distributed medicines and more particularly food. Most of it had been lost during those terrible days. They ran a rationing service by helicopter: so much drinking water each, two *pannini* (those little pointed loaves about five inches long) for each adult each day, a little flour, one and a half ounces of sugar for each adult, three ounces for children, no milk, no coffee. At one time they had a population of 65,000 inhabitants of isolated and often devastated villages to care for. As communications were re-established the news of casualties mounted steadily: Venice 4, Belluno 41, Trento 26, Udine 14, Verona 1, Vicenza 3. The final total was 71 dead, 120 injured. In the official list of damage for the whole province of the Veneto four towns and their administrative areas were listed as gravely damaged (Bolzano, Grosseto, Trento and Venice) and nine others as having extensive damage. In all 492 communes were damaged by the action of the sea, the rain and the wind.

By November 24 we had launched a UNESCO General Conference in Paris. My first hint of the disaster was the stricken look of the Italian delegation. There were a number of Venetians and Florentines amongst them, deeply preoccupied about the safety of their wives and children and the fate of their native cities. The Italian Embassy in Paris was in constant touch with Rome on their behalf and everything possible was done to obtain infor-mation for them. Necessarily, in the conditions prevailing in both Venice and Florence, it took time to get news. It was an anxious time for them but eventually all of them received messages, brief but comforting, reporting the safety of their families. I only had a house to worry about and could not possibly ask for news. I decided that the only thing I could do was to leave for Venice myself as soon as I could, unannounced, as all my friends and advisers would have their hands quite full with their own prob-lems. On November 18 I managed to get away for the weekend with a pocket full of messages for the families of friends, in the event that the telephone system was back in commission. There were many pleasant surprises in store for me that weekend. The first was that the train was running on time. Railway traffic between Milan and Venice had been re-opened on November 12, all the culverts and bridges having been, at least temporarily, repaired, and the line everywhere checked for safety. On my arrival the station seemed quite normally clean and the coffee bar as usual open and bustling with customers. It was with some apprehension that I walked down the steps towards the Grand Canal but again everything seemed in order. A superb spring cleaning had been done. Practically every trace of the flood had been removed. The speed and effectiveness of this operation must be measured by the fact that hundreds of shop windows had been smashed; that hundreds of doors had been broken down by the floods; that many cisterns containing fuel oil for central heating boilers had burst spreading a thick greasy black coating over the paving stones and walls of so many *calli*. But all traces of these things had disappeared. An occasional smear of fuel oil on some inaccessible wall of a back canal was all I saw. My worst moment was in the *vaporetto* coming round the corner from Mazzorbo at the point from which one can first see the house. I let out a sigh of relief. There it was. At least it was still standing. As we came in to the landing stage a local man whom I hardly knew shouted to me, '*Tutto va bene, signorina.*' The Italians are really astonishing people, placing themselves in one's shoes with a quite extraordinary rapidity, imagination and kindness. I shook hands and thanked him and found Giulindo working in the vineyard. He confirmed that, indeed, everything at the house was quite all right. My excellent architect had taken precautions against floods and had built the house so that the floor level was forty centimetres above the highest-known flood mark for the last two hundred years. The water had come to within four centimetres of the ground floor level. Both the back and the front terraces had been awash. No water had come in. The house was damp, since the water had come higher than the damp course, but safe. It had been so well built that the gales had had no effect. Not a single tile had budged, not a single shutter had blown open.

The vineyard, on the other hand, looked like a battlefield, littered with tree trunks, rubbish and bales of straw. My builder had breached the embankment to bring in his equipment and materials and had, contrary to his instructions, left a yawning gap in the protecting bank. I suspected that he had not rebuilt the bank and, in Paris, every time I thought of that gaping hole I had a nightmare vision of a rip-tide hurtling the marvellous, fertile soil mixed with uprooted vines and young fruit trees through the breach into the lagoon. The sea had, indeed, come flooding in through the gap but forty-eight hours later had gone gently out again. In fact, thanks to my builder's incompetence, I was much better off than my neighbours whose land lay sea-water-logged for days. They had finally to blow up their banks to get rid of the sea-water and when their land had dried out, they had been put to the expense of rebuilding them. It was difficult to judge the extent to which my young vines had suffered from being soaked in salt water for two days since they are dormant in November. In this part of Italy the vines grow on stocks some four to five feet tall clothed in a rough, thick bark. The new branches which each year bear the grapes spring from the top of this stock. Most of the mature vines in the island had kept their heads out of the salt water and were expected to survive. So we had an anxious wait until the spring to see how many of the young stock made buds. In fact we only lost about 150. I was so relieved that I came to regard this as a great blessing. Artichokes, on the other hand, will not tolerate salt water and everybody in Torcello had, once again, lost all their artichokes. The tractor was another casualty. The effect of lying for two days in salt water was to develop in it a tubercular cough just like the heroine of a romantic novel and, after a few months, it had to be replaced.

As the ground dried out and the warmer weather of spring approached it was possible, on any sunny day, to see a sheen of gleaming salt over the soil. I remembered the great floods which invaded eastern England in 1953 and the care which had to be taken to rehabilitate the soil over which the sea had swept. I sought the advice of the Ministry of Agriculture in London and they were most helpful in sending me reports of their experience with the use of gypsum to counter the effect of salt on the soil. But in Italy the cost of the gypsum and the transport and the labour for spreading it would have been enormous. So, like my neighbours, I had to wait with what resignation I could muster until 'the earth's inward narrow crooked lanes did purge sea water's fretful salt away'. Some said it would take two years and some five. In fact we seemed to be able to grow almost any crop after two years.

The flood was a major catastrophe in the lagoon. Coming after a hard winter and a summer of unsaleable profusion, many families were at the end of their resources. There were hungry people in the lagoon that winter. The Government did what it could but Italy is not a rich country and the grants to replace cattle and vines were necessarily minimal and, in any case, only time could make them productive. The rich northern Italian plains are in no way comparable to the infertile south. Yet one saw in the lagoon that winter something of the same terrible resignation of peasant people that Carlo Levi so starkly describes in *Christ stopped at Eboli*.

Unlike the Florentines, the Venetians are prepared mentally for floods, and a tide—even a flood tide with the *scirocco* behind it— does not rage quite as the waters did through the streets of Flor' ence. Nevertheless the Venetian flood had a traumatic effect on the citizens. Their terrible experience had borne in on them with a new vividness how vulnerable was their city to the senseless fury of the wind and sea. Many of them thought with a new sense of horror how unsubstantial, encrusted, almost fossilized were the oaken piles on which their fragile city stands against the onslaught of the water, and how dangerous was the equally insidious seep of humidity. For the first time, it seemed the mass of Venetians questioned the capacity of their city to outlast time and talked of its future with a harassed and tender anxiety. This was not a passing mood. Early in 1970 the city was plastered with posters about the need to save Venice. Shops and cafés were asked to close for the afternoon and evening to mark their anxiety about their city. A foreigner who was there told me that it was a city in mourning.

As soon as order was restored in the city and the province, as soon as their dead were buried and repairs in train and communi/cations re/established, they began to take stock.

The problem of what must be done to protect Venice falls into three broad interlocking divisions. There is the need for scientific enquiries to discover how far it is possible to predict dangerous changes in the oscillation of the Adriatic (similar to the early warning system devised for hurricanes in the Atlantic) and the consequent engineering work required to protect the lagoon from it effects; to throw more light on the causes of the subsidence of the bed of the lagoon, which appears to be one of the reasons for more frequent flooding; and to determine the extent to which pollution of the air and the lagoon is a danger to the city, and suggesting consequent action. The next is action to restore buildings and objects which have made Venice famous. The third is the need to study the social action required so that Venice may continue to be a living community. Over all these hovers the need for money, for these three requirements, without which Venice will not survive, are all enormously costly.

Finance and responsibility are closely linked. The main respon/sibility for Venice must lie with the Italian Government and with the Venetians themselves. We all of us non/Venetians, whoever we may be, who have a stake of affection or admiration for Venice, feel that in a sense it belongs to us. The current fashionable theories that certain peaks of artistic creation in the world are part of the cultural heritage of mankind can obscure the more important fact that they are the products of a particular civilization, that they are a mark or badge of its identity. It is not reasonable to expect the same kind of affection, admiration and understanding for the products of another civilization as one has for one's own. One can—one certainly should—greet them as being in the same cate/gory and recognize that they evoke the same kinds of emotion in members of the civilization which created them. But a Brown is not a Jones and, in a sense, it is to diminish Brown to pretend that he reacts in the same way to the same stimuli as a Jones. Why should he fuse his identity with Jones and why, as they are marks of his identity, should he not prefer Brownish things and even understand them better and so be fonder of them than he is of Jonesish things? Venice is important to some of us of Western European civilization because we have the same roots, because it is recognizably of our civilization, while being also recognizably of a particular Venetian and Italian branch. This lays on us also a responsibility for Venice: a secondary responsibility. The main responsibility, particularly for decisions about what should be done, must rest with the people who created it. It would cease to be Venice otherwise.

The Italian government has taken its responsibility immediately and imaginatively. Parliament has before it a Bill for the enormous sum of 400 million dollars (about £170 million) for various scientific, restoration and social rehabilitation needs in Venice. When one thinks of the many things which need to be done in Italy which, in spite of the economic miracle that has made the North so prosperous, is not rich as we are rich, one must recognize the far-sighted, imaginative spirit which moved the government when it proposed to spend this large sum on one city. When one thinks of the many other cities of Italy whose monuments are crumbling, whose pictures and statuary need restoring and whose social conditions are as old-fashioned and restrictive as those of Venice, one must surely admire the self-sacrifice which is demanded of the deputies of other cities to devote this money to Venice. One must hope that nothing frustrates the intentions of the government and that this huge sum will be safely voted. Venice itself, its municipality, its institutions, whether they are branches of a state or of a local service (such as the Magistratura alle Acque, which oversees all matters relating to the lagoon, or the Soprintendenza ai Monumenti, which is in charge of the preservation of monuments and sites), its learned societies (such as the Venetian Institute for Science, Literature and the Arts) and associations of well-disposed, and sometimes very influential, individuals (such as Italia Nostra) have also taken their responsibilities rapidly and imaginatively. So far there are almost as many plans as there are Venetians and the most difficult problem facing them may well be to find a way of sinking their differences whether they arise from scientific, professional or political reasons. This should not be beyond them. Venice is a man-made city to a greater extent than others. It arose from its improbable first beginnings on a mudbank because its citizens worked together. Lacking any natural wealth, it became rich and powerful on the resourcefulness in co-operation of its merchants, bankers and sailors. The modern Venetian owes it to himself to return to this tradition of common effort to ensure the resurrection of his city. He must recognize that an agreed plan has now become a pressing necessity.

There is a place amongst this complex of requirements for international action and this, too, is being filled at this stage chiefly by the Council of Europe and by UNESCO. Their first contribution was to make the dangers to Venice widely known and in this they have done well. The second was to attract the co-operation of scientists and technicians in all the specialities required on matters as diverse as sea currents and subsidence, the building of locks to regulate the flow of water between the lagoon and the sea, the preservation of stone and the restoration of works of art. The third will be to stimulate and possibly to act as a channel of assistance from governments and private sources outside Italy. This seems to be proceeding slowly in spite of work already being done by some governments, particularly the French; and, to quote examples from Britain, institutions such as the Victoria and Albert Museum which has sent out experts in restoration, to work with Italian colleagues in Venice, and by the 'Venice in Peril' fund, which under the chairmanship of Lord Norwich, is very actively collecting funds.

There is a great need for international assistance. On the scientific and technological side there has, so far, been no great call for large sums of money because research must first be carried out to make sure what should be done. But eventually there are bound to be large civil engineering requirements to protect the city and the lagoon from the action of the sea, from industrial effluents

from the factories and oil refineries sited along the lagoon and
from the need to make reasonable and convenient access between
them and the sea. The damage to buildings and works of art
arises chiefly from harm done by humidity, the salt and acids in
the atmosphere and, in the case of external walls, the waves
created by the increase in fast motor-powered traffic on the canals.
The scope of the work required may be seen from the schedule
established by the Soprintendenza, with some financial assistance
from UNESCO. This lists 392 palaces and town houses which
require immediate attention out of a total of 700 needing some
repair; 126 churches in Venice or the lagoon islands requiring
some restoration work, as well as twenty *scuole*, which are the halls
of ancient charitable guilds. To this must be added work on a
similar scale required for the restoration and preservation of works
of art and precious objects in museums, galleries and churches as
well as books, manuscripts and documents in the many libraries
and archives of the city. Again, the main cause of damage is
humidity and acid from the air. Frescoes suffer particularly from
these conditions though many detachable paintings are scarcely
less vulnerable and present difficult problems of conservation,
some being very large and others so incorporated into the
architectural decoration of the churches as to be practically impossible to remove for care and cleaning.

The main cost of this enormous effort will however undoubtedly
be for the social action required to maintain Venice as a living,
active city which can take its place on equal terms with other
cities of today. This involves ensuring work, improving housing,
which is expensive and difficult in a densely populated area,
particularly when built on unstable foundations and where by far
the greater number live in massively built tall old houses; providing amenities and transport to meet contemporary expectations.
Amongst the improvements suggested for transport is an underground railway system for the city which would emerge to stride
across the lagoon linking Venice to the Lido, and the main
islands with Treporti at one end of the line and Mestre at the
other. Less radical, but likely to provide a solution with less

damage to the city, is the proposal to build underwater garages (like the one below the lake in Geneva) off the Fondamente Nuove with access roads from the present causeway linking Venice to Mestre. Amongst other projected amenities is the new hospital to be built on the basis of plans originally drawn by Le Corbusier. This long-delayed project would add notably to the architectural interest of the city. Venice is sadly lacking in distinguished examples of contemporary style in architecture. There is a contro-versial project by Wright for a new *palazzo* on the Grand Canal to house university students which on paper looks exciting, but which is likely to be abandoned under pressure from the ultra-conservative, and a new bank by Nervi which so far seems to have escaped equally bitter attacks, perhaps because it is on a *campo* near Rialto and not in so visually sacrosanct a situation as the Grand Canal.

From the destruction of the ancient republic by Napoleon to the Second World War Venice was a poor city, vegetating in its lagoon with little industry, apart from its port and its tourist trade, and a mountainous and, on the whole, economically backward hinterland. After the war it began to stride forward but a number of factors prevented it from making a wild rush into modernity. In the first place, while no longer so poor, Venice was still by no means rich and could not afford to take advantage of processes which were now technologically possible. In the second place, the Venetian, while admiring (and often yearning after) modern-ity, is basically a conservative, particularly in spending his money. Consequently, in spite of pressures which built up within the city over the years, no action was taken of the kind that other richer, more active cities have learnt to regret. From the period soon after the war, when technical processes developed for military reasons were applied to civilian purposes, until quite recently, every advance in science and technology was welcomed and exploited for its own sake—and for the benefits which it could bring to mankind in health, convenience or amenity. It did not appear to enter anyone's head at the time that there might be drawbacks to these new processes. In England when detergents were first used in industry there was an outcry against the unsightly scum floating down the streams into which the used water was pumped. No-body thought that there might be other ill effects. This compart-mentalized thinking, this habit of looking at one thing at a time, of ignoring the fact that *tout se tient*, that everything affects every-thing else, led to decisions from which we are likely to suffer for many years. Barry Commoner recently said in discussing what the Americans (who had the money and the energy to exploit new methods immediately) had done to their environment: 'Once you understand the problem you find that it is worse than you ever expected.' This same compartmentalized thinking does not only affect the rich. It applied in the case of the Aswan High Dam and is producing grave adverse effects in Egypt. Water-weeds have so increased in the dam and have created such extensive evaporation that sometimes there is not enough water to drive the generators. On the other hand the rich alluvial silt of the Nile no longer reaches the sea and consequently the anchovy catch declined from 18,000 tons in 1965 to 500 tons in 1968. This kind of calamity has not overtaken Venice. Pollution is there and is increasing but has not until very recently existed on the scale that is met with in many other large towns of Europe and North America. Venice has been protected by its relative poverty and its geographical isolation though it has of course suffered, as we all have, from other people's decisions which we have none of us, so far, found a way to control. The indestructible plastic container which lurks under every fence in all our countries and, in Venice, bobs for ever in the small canals and backwaters, is a case in point.

Venice has been fortunate in that a recognition of her danger has come at a time when the world has become sensitive about the environment. This means that Venice should be able to benefit immediately in a practical way from the research and experience of other cities of the world. The effects, for instance, of the pollution of the air on health as well as on stone buildings and metal is now well documented and, in consequence, the application of this experience to the Venetian situation should be more rapid. In a similar way the main dangers of water pollution on flora and fauna as well as on health and amenity have been extensively studied. It is now known that factories can remove deleterious elements, which harm the surrounding areas and the people who live in them, from the fumes which they pour into the atmosphere and the water which they pump back into the rivers and seas after they have used it in industrial processes. It costs the factories money to clean the air and water they have used but, as our own clean air areas in London and great manufacturing centres so clearly indicate, there is an enormous benefit to be derived in making them do so, even if we have ultimately to pay the cost in higher prices. So, Venice's need to protect herself and her citizens and to keep in balance the ecology of her surroundings is likely to fall on sympathetic ears. This new interest in the environment should, in turn, be some help in encouraging governments to assist the Venetians and the Italians in the enormous task they have taken on in halting the rotting-away of this ancient and beautiful city.

Venice has perhaps also been fortunate in being a static, if not a stagnant, society during the last few generations. During this time the surrounding water has put obstacles in the way of pursuing two courses which other cities now bitterly regret. The first is the growth of suburbs to keep pace with the growth of populations, and the second is the need to accommodate the motor car.

Venice, of course, has Mestre. But this is a suburb with a difference: it is isolated from the centre by a stretch of water. Because of its physical situation Venice could never be subject to such a monstrous growth as Greater London, or the huge Californian subtopia with city nudging city all along the coastline from San Francisco to Los Angeles, or a milling hive like Tokyo or the extensive shanty towns found in developing countries such as Calcutta or Lagos. Nor does Mestre seem to create those social and psychological difficulties which the combination of a great town and dormitory area seems to produce in other places such as the London suburbs or even the new tall dormitory blocks growing up around Paris. In spite of all its manifest drawbacks, it remains more alive than these, perhaps because it is a place where people work as well as sleep.

Venice has also escaped the motor car, though regretfully. To the Venetians it is a symbol of the twentieth century and they pursue it with passion. To them it represents the speed and mobility which is the mark of contemporary man. Without it they feel frustrated of a legitimate desire to be like other men who live in happier communities unimpeded by water. There are wild men who would fill in the Grand Canal, turn the Piazza into a parking lot and run a highway along the Fondamente Nuove and across the water to the Lido and other lagoon settlements. But Venice did not have the money to put such schemes into operation even if the majority had concurred in them. A great many Venetians possess cars, perhaps more than possess boats. On Sunday evening in the summer there are long queues rivalling those in any large city of the West crawling and stopping, crawling and stopping all along the road leading from the resort towns and beaches between Venice and Trieste. At the same time the lagoon, though bespattered with outboard motors, is comfortably empty.

Other towns are now taking measures to keep the motor car in its place, to reduce it to the role of convenience and pleasure that it undoubtedly has, but to prevent it becoming a tyrant to distort and ruin our lives. Rome has forbidden cars to enter the Piazza Navona and that elegant circus now lives and breathes and rings again to human footsteps. Rouen has closed the rue du Gros Horloge to traffic for certain hours of the day. Rotterdam was rebuilt with shopping areas accessible only to pedestrians and the the same pattern has been followed in some of our own new towns. In New York a section of Fifth Avenue is closed to traffic at weekends and motor cars are forbidden on public holidays and weekends in the centre of Tokyo. Paris is still largely under the spell of the motor car. It is true that they have been ousted from the Place Vendôme recently but the lower quays along the Seine, which used to be the quiet resort of fishermen, lovers, students with their text books, Sunday painters, impecunious tourists with their modest sandwich and the lunchtime clerk reading his newspaper in the sun, have been given over to an endless stream of speeding cars. The increasing number of underground garages and under passes presents another danger to this marvellous city which was, until recently, a paradise to walk about in. So many trees, which have survived being stifled by petrol fumes, are being felled to accommodate underground parking places. Young trees are mostly being replanted but even if they survive in the shallow soil laid over a concrete and rubble base, two or three generations will be deprived of their shade and elegance, and what will Paris be in May without the chestnut trees flaunting their white and pink candles? London, alone of the great cities, seems to have made no compromises so far. Certainly I do not seem to feel deprived in London, as I do in Paris or Rome, of space to walk. Perhaps this is because the centre of London is set about with great parks— St. James's, the Green Park, Hyde Park, Kensington Gardens and the Regent's Park. There are, alas, no longer any grazing sheep in these parks but there are trees and grass. Pedestrian sanctuaries—and how far the evil must have gone to describe them in such a phrase—bring back a liveliness to the streets which such places as the Calle de la Sierpes in Seville and the *calli* of Venice have never lost.

If it is true that cities are the accelerators of social change Venice, imaginatively rejuvenated by its citizens, could have much to teach us all about an urban environment. But they must first look to their own situation. Venice is old and dilapidated in the physical sense that its buildings and plumbing are out-of-date and out of repair, but much more important is the fact that its social attitudes are out-of-date. It is a city very conscious of class division and property, and the people who possess these build up defences around them. This has always been so in Venice. The Serene Republic had rigid social subdivisions. It was practically impossible to move from one social stratum to another. This eventually created the stagnation which is undoubtedly one of the reasons for the city's decline. The unification of Italy was a nineteenth-century achievement at a time when the middle classes were in the ascendant throughout Europe, and Venice since that time has settled down into a new but again too rigid a pattern. Venice is a very comfortable town for the middle classes to live in, though they are beginning to complain about the shortage and expense of servants, but it is not a pleasant place to live in for the working people. This is one reason why they are going to Mestre. Venice is, as a result, in some danger of becoming a middle class centre and so losing the vitality which comes from all types of people, doing all kinds of jobs, living and working together. One of the many things wrong with suburbs is that they are socially speaking too homo-geneous and it is this that gives such formidable strength to the social pressures to conform to a pattern which eventually stifles the society that lives in them. The Venetian municipality has in train a whole plan of regeneration of housing in the poorer parts of Venice and large sums of the money ear-marked by the Italian Government will be devoted to the purpose. But to weld the com-munity together needs more than improved housing: it needs a change of attitude. Why, for instance, should a member of the pro-fessional classes not hesitate to address a peasant by the familiar '*Tu*' when he would undoubtedly use the formal '*Lei*' to a fellow pro-fessional? And, of course, this kind of attitude exists in reverse. I suspect that this split between the working people and the bour-geoisie is one of the main problems facing Italy, as it does France to at least an equal degree, and which is only slowly being pushed back in England. It takes time to change attitudes and to shift habits but it must come eventually if the society of Venice is to match the harmony of its setting. In the meantime more must be done, as is planned, to provide amenities for the poorer people and especially for the poorer young people. Life is really very dull for them. It is about as dreary as it was in the nineteenth century. It is true that television has brought them much relief but it seems to have left many of them with an unsatisfied feeling, a vague idea that life is gayer and more dashing and exciting in other cities not surrounded by water, where you can every night take your girl out on a motor-bike and dance to exciting new music in exciting new places. Life is, of course, not like this but if you are as bored as the young people in Venice are bored it is the fact that you think it is like this which matters. The problem of young people in Venice is no different from that which faces any large urban population. Perhaps here, our experience in youth centres in our big cities could be relevant to the Venetian situation. A self-governing community running a centre which can provide scope for a wide range of interest, under the tactful and unobtrusive guidance of a youth leader, would do much to release energies and assuage grievances. It is amenities for the winter months which are particularly needed and there is not much difference between a winter in the more congested part of Leeds and similar quarters of Venice. Places are required where young people can meet and play games and buy a coffee, dance and talk or play music, or maybe swim in a modern indoor pool.

Venice needs to exploit its advantages imaginatively. A place like Venice should rank with Oslo in its reputation in water sports. There is a famous rowing club and local regattas are well attended but what has been done to encourage dinghy sailing, or water skiing or water polo? If one looks across the Adriatic to Yugoslavia one finds flotillas of dinghies belonging to youth clubs putting out from every port. Some Venetians tend to be a little plaintive about the lack, for instance, of a football ground for their young men. Soccer is the guiding beacon of many lives but more people in Leeds and Glasgow or Lille or Düsseldorf have to go further and take more time on the journey than a Venetian to watch a match at S. Elena or to play himself somewhere on the *terra firma*.

The Venetians need to resist the feeling that they are old-fashioned, that all the gadgetry of the twentieth century has left them behind. Because of their geographical position they have been backward in exploiting certain products and techniques of scientific progress. The motor car is the most obvious example. But this fortuitous backwardness has served them in good stead because Venice is one of the few cities of the world that has not ruined its environment to accommodate it. It is still a city built to a human scale. The population aimed at in the new plans is about 150,000 inhabitants. It will at all times be possible to go from the centre to the edge of it within half an hour or so. And it will be possible to go on foot. This brings life back to a human tempo, avoiding those pressures which built up to the 'rat race' so many Americans are now seeking to avoid, or to that dangerous overstrain which the French can only parry by being declared *fatigué* by their doctors and ordered away for four weeks in the mountains. And how many Englishmen die in the first year of their retirement, when the strain lets up?

Venice has a great problem before her. But she has all the elements for a solution in her own hands. Because of her recent backwardness she can leap over the mess that other cities have made in this century straight into the twenty-first century, and in doing so she can teach us how to repair the damage we have done to the lives of our cities. If the Venetians can find an imaginative solution to the problem of building a harmonious society, in a city designed to a human scale, exploiting all the advances and conveniences offered by technology and protecting the environment and the needs of other things that grow and live without which, we are beginning to realize, we cannot continue to grow and live ourselves, Venice will have done more for humanity than ever she did when she founded a proud republic or built one of the most beautiful cities on earth.

CHAPTER XI

THE FINISHED HOUSE AND THE 'TERRA FIRMA'

By Easter 1967 the house was all but completed inside and out. There remained the fiddling process of minor adjustments; a door that did not hang quite straight; a radiator that was set askew; some shutter catches that were too high and some too low and so on. The *capo di lavoro* concerned came with great amiability to make notes, converse, sip coffee 'corrected' with *grappa* and go away never to reappear. I spent my life trudging along to telephone from the little bar to plead with them to come back but they were always busy building a house for somebody else. Slowly I realized that this maddening behaviour was not due so much to inefficiency, except in the case of the plumber as I was beginning to discover, as to a quite genuine lack of interest. Building my house had been an interesting job, and a challenge in the same sense that a *sistemazione* presented a challenge. Now the job had been done and they were interested in something else which pre-sented more of a challenge than the correction of minor details. By a calculated mixture of cajolery and fits of temper, gradually things were made good, but the process lasted a very long time. Nor was I much comforted by my architect's comment that five years after completing his own house, there were still minor adjust-ments to make, if only he could get the *capo di lavoro* to do them. I turned my mind away from these difficulties as much as possible because I realized that they presented a long grind and that the victory would go to the most persistent. There seemed little point in expending nervous energy on such an uncreative, indeed sterile, process as insisting that the hot water supply produce hot water.

I now learnt about the constant vigilance required to maintain a house. In the lagoon we are subject to both the blistering Medi-terranean sun and to the insidious attacks of humidity and a salt-laden atmosphere. The result is rather like living on a ship at sea. Every day, every day one goes round with an oil can in one hand and a paint brush in the other maintaining intact the defences against heat, humidity and salt.

The time had come to think about furnishing. I did not expect that Torcello would become my permanent base for some years. At that time I would transfer from Paris a whole house full of furniture most of which was English in style. My problem was to furnish the Ca' S. Tomà with pieces which would not duplicate what I already possessed but which would provide the basic needs until I brought the rest of my furniture from Paris. The bedroom presented no problems as adequate space for clothes was provided in the built-in cupboards, and the beds I already possessed were of no particular interest or especial comfort and could well be sold off when I left Paris.

The beds were the first indication of the amusement to be derived from following wise local help. It was my architect who told me first that in Venice one did not buy things but had them made to order. In this way, said he, one was sure of good quality at a lower cost. As a result I interviewed an upholsterer working on his own account whom he had recommended and one day, when I arrived at the house, I found five admirable handmade beds with both spring base and mattress of horsehair in stout damasked covers with fine fat feather pillows to match at a cost of fifth-rate factory-made beds in England. Much encouraged I set off to interview a relative of Giulindo's with a great reputation as a cabinet-maker. He has a little house and workshop beside a canal in the fields near Treporti and when I arrived he was busy making a rowing boat for a customer which he told me he was doing for fun as it was not really his trade. I wanted a kitchen cupboard with an old-fashioned ridged wooden draining board made to fit on to the sink. I had brought all the measurements with me and could not for some time imagine why he was looking so startled and dubious. It then turned out that a draining board such as I described was something he had never seen. So I found myself, several years after beginning the planning of the house, still caught out by my national attitudes to everyday things which are, of course, the ones that one most easily takes for granted. On the other hand I had the pleasure of seeing a good craftsman warm to an idea. Here was another Italian caught by the challenge of a *sistemazione*. The draining board which he made is a singular affair to an English eye but it works very well. It is not ribbed and is made of a gleaming white formica which he decided would be easier to keep clean.

I soon discovered that in the hill towns behind Venice there is a strong tradition of craftsmanship and that there exist numbers of cabinet-makers and upholsterers working on their own behalf. There are also a number of small factories making good, simple country-designed pieces, and it was in one of these factories situated in a village beyond Novi, a centre of famous but now mostly repulsively ugly ceramics, that I bought some plain and solid oak rush-seated dining-room chairs which are very pleasant in their rustic way. Though I was not yet ready to place many orders I took the opportunity to call on a large number of crafts-men to see their work. It was in this way that I met an excellent cabinet-maker who lives just outside Asolo. His little house stands in a lane and the hillside falls away so abruptly that his workshop nestles under the house and gives on to the orchard. Here he talked about his work and showed me what he was making while his children chased one another amongst the shav-ings under the work-bench. He liked to be given precise, measured working drawings and was ready to make anything from them. He had a stock of seasoned wood and was prepared to look out for old wood or make up my own old wood if I had been for-tunate enough to find any. He did not like to be hustled and told me that he would like six months' notice to complete a large cupboard, or a Renaissance-type refectory table. Talking with him in his workshop with the sun pouring through the window and the meadow, strewn with flowers, sloping down the hill, and his wife with her tub under a tree thumping her washing, I thought of the artisan's England described in George Bourne's *The Wheelwright's Shop* which was already disappearing at the begin-ning of the century. Here was the same pride in a traditional craft surviving in Italy without selfconsciousness and quite naturally allied with an old lorry for fetching wood and delivering his finished work parked under a trellised vine and a television set in the living room.

It would be wrong to give the impression that all Italian artisans working on their own account are fine craftsmen. There are some who waste their skill on poor materials and some who are content with botched work if they think their customers will accept it. My plumber is a case in point. But this is not to deny that there is a strong tradition of craftsmanship and, as there is still a preference for hand-made goods, there are, for the time being at least, suffi-cient customers who prefer to go to an artisan working on his own behalf to keep him in business. It is a fact that the furniture industry in Italy is still a fairly new one, and, at present, there does not seem to be much difference in price between well-made factory furniture and what a good artisan will charge. In England the difference in price is very considerable and we might with the money buy antiques. Perhaps that is why we do. Perhaps because we were the first industrialized country, we long ago lost the habit of ordering things to be made for us rather than going to a shop which has the attraction of a choice of factory-made goods. Yet the success of those few furniture shops which pride themselves on design and craftsmanship would lead one to assume that there is room for the artisan in reaction, maybe, to industrial products. In France there is a definite revival of interest in craftsmanship out-side those trades, such as food, wine and clothes, which have never lost their craft methods. Perhaps it is because of these trades that it has been easier for the French to make that side-step into an appreciation of craftsmanship in other trades. The tendency is no doubt assisted by the fact that everything is so expensive in France that one might as well pay rather more for the satisfaction of obtaining a hand-made article. It would, in any case, be pleasant to think that in future the customer would be able to choose between buying a well-designed and well-made factory product at a reasonable price or putting rather more of his money into a hand-made one, and equally that a skilled man would be able to choose whether to exercise his craft in a factory or be sure of making a living if he launched out on his own account. The choice is more likely to be present in a country which has pre-served a tradition of ordering things from an artisan, than one, like ours, which has almost lost most hand-skills.

These journeys to consult craftsmen provided opportunities of revisiting some of the hill towns and villas and of discovering new enchantments in this region. Asolo is one of the high points of the Veneto with its castle, its memories of its romantic queen and its frescoed houses. It is crowded with visitors throughout the season except for the little streets winding upwards to the top of the town and the walks in the folds of the little hills where primroses and violets carpet the banks under the trees. To an English eye there are many resemblances to our countryside of deep lanes and little woods and I always feel that it should have been during his stay in Asolo that Browning wrote about the wise thrush which sings each song twice over.

The fortified towns of the *terra firma* come as a shock after the wide-windowed indefensible *palazzi* of Venice which emphasize both confidence in the value of the sea as a rampart against the enemy and recognition that if the enemy reached Venice there was no defence possible. These fortified cities are also a reminder that Venice held the *terra firma* by the strength of its armies. Castel-franco, with its pink walls and four-square military layout, is one of the most attractive of these towns. Marostica, with its crenellated walls marching up the hillside like those other walls of that other Venetian outpost on the island now known as Hvar in Yugo-slavia, remind one that Venice maintained colonies and fleets to protect her trading stations on many coasts and islands of the eastern Mediterranean. Further away from Venice are little towns which have almost the air of provincial capitals. Udine is such a town with its palaces and its clock tower recalling that of Venice and the lion of St. Mark stamping it all as Venetian property and not independent at all.

The *terra firma* has throughout history served as both the play-ground and the home farm of Venetian patricians. One of the astonishing things in the villas which are found at every turning in the Veneto is that even as early as the fifteenth century they have had the high broad central windows similar to those in so many Venetian palaces. The Collconi Villa at Thiene is a very good example. Villas dating from this period seem to have been sur-rounded by a crenellated wall but they do not appear to have resembled the grim fortress-like structures that our ancestors were building for themselves during those years in England, where confidence in one's neighbours did not come till very much later. The Venetians' confidence was indeed a little premature. Many of these villas were destroyed in one or other of the savage wars that the Italian cities waged against one another. But two generations later Palladio was building his villas without any provision for defence and with wide windows and elongated wings designed for commodity and proportion rather than with an eye to fortification.

The Villa Barbaro at Maser, some two miles from Asolo, is one of the most gracious of these villas. It stands almost at the top of a gentle slope and two long arcades, decorated at the end by simple-shaped gables, are separated in the centre by the house which juts forward under a pedimented roof. Projecting above the long roof of the house are dark cypresses and the other trees mostly of a sombre foliage which throw into relief the pale yellowish terra-cotta of the walls and pillars of the arcade till it glows between the sombre trees and the green of the sloping lawn before the house. The elegance of the decoration inside matches the superb exterior, particularly in the saloons for which Paolo Veronese painted frescoes of a charming domestic kind: a lady smiling over a balcony at her guests assembled below, her page playing with his monkey a few paces off; the buxom maid entering through a painted door at one end of the series of intimate sitting-rooms while a huntsman with his dog and his gun negotiate a door at the other end. The wide arcades are said to have served as farm stores but at a fairly early stage the smell and din of the farmyard seems to have been banished to the 'factory' (*fattoria*) which is a great barn built a little way from the house for the storage of produce and implements as well as housing the farm servants. At Maser this building stands some two hundred yards away on the same contour of the slope. At the Villa Foscari—or the Malcontenta as it is usually called from a story that a cantankerous wife was banished there by her family—which is the nearest to Venice of the great Palladian villas, the *fattoria* stands right beside the house. The Malcontenta is best known from its pedimented façade facing the Brenta canal, framed by huge weeping willows. Inside there are on the main floor a series of rooms grouped round a central saloon, making a ring of interconnecting reception rooms. Palladio pushed this development to its logical conclusion at the Villa Rotonda built on a hill on the outskirts of Vicenza. It was

planned not so much as a house to live in as a place for elegant
entertainments and routs, and it was perhaps with this purpose in
mind that he so sited it as to provide a different and harmonious
view from each of the four stately façades. Lord Burlington (who
was largely responsible for the wave of Palladian building
which influenced the great English country houses of the eighteenth century) was so taken with the Villa Rotonda that he followed its design very closely in the villa he built for himself at
Chiswick. The Villa Rotonda is now normally closed to the
public but any visitor to Vicenza should climb the little hills
where it stands, partly to see the Villa Valmarana ai Nani with
elegant chinoiserie decoration by Domenico Tiepolo, but also to
enjoy the surrounding slopes and little hidden valleys which display the works of agricultural man at his most harmonious: one
expects to find the shepherds of almost any pastoral lay piping to
their sheep under the shade of a tree.

The country villas of the Venetian patricians were mostly for
enjoyment. Some, and amongst them the greatest, are, like Maser
and the Malcontenta, isolated in the countryside. But if the owners
sought health and coolness on the *terra firma* they were not, for the
greater part, seeking a rural retreat. They continued to live a
sprightly social life. All along the banks of the Brenta canal there
exists to this day a splendid array of villas great and small. Similarly
the outskirts of Treviso (now a bustling market town) in the direction of Venice where the road is lined with villas standing in large
gardens decked with evergreen magnolias and planted with shady
walks. The great villas were also the centres of large agricultural
estates, for the Venetian patrician, with his banking interests and
ships at sea, was unlikely to leave an estate lying idle. It is strange
to find that even today so intensely urban a people as the Venetians
should so often have a foot in the countryside. It is not odd that the
remaining patricians, however much diminished from their
former wealth, should retain a country house. What was, to me,
unexpected was to find so many middle class people with a house,
and a little vineyard and orchard somewhere in the Veneto. Now
the motor car and the fashion for mountains take the Venetians

further afield and often up into the Dolomites. Venice can become very stuffy in the summer. The Lido and the islands of the lagoon usually enjoy a slight movement in the air, which is not quite a wind but which tempers the heat. The moment the land rises a little as, for instance, in the hills where Asolo stands, the sultriness goes out of the air. In the real dog days the best escape is to the Dolomites. A three-hour drive from Venice takes one up to the pine forests at such places as S. Martino di Castrozza crouching at the foot of the extraordinary pink peaks of the Dolomites.

I used to escape up here sometimes when the heat of the plains diluted my enthusiasm for consulting with craftsmen or trying to discover where to buy wine. When I came to live in France I had the great good fortune to be introduced by a friend to the habit of buying wine direct from small *vignerons*. It has since become one of my greatest pleasures to seek them out and to collect my wine by car. In the first place a well-tended countryside in itself is a pleasure to visit. The neatly staked, well-weeded rows of vines stretching away up the slopes are often confined, at least in Burgundy, by sturdy stone walls. It is also a pleasure to talk to the highly skilled men who tend them. The growing and making of wine is a highly scientific branch of agriculture but it remains also an artisan activity, for what makes the difference in the final analysis is the skill of the man in charge of the process. There is no more delightful way of spending a Sunday morning than leaning against a barrel with a silver *tastevin* in one's hand discussing the pros and cons of a wine with the man who planted the vines and pruned and cosseted them and finally made, with all his intelligence and experience, the marvellous wine in these great dark *fûts*. I remember once being invited to go down to Santenay to taste some wine from a small, choice, vineyard. When we arrived in the village it was, like so many French villages, empty and the houses were close-shuttered as against an enemy. They must have been on the look-out for us because as we drew up the gate flew open and we were ushered into the yard. On the side parallel with the street was a single-storey house with the cellars below. On the other side was a big barn-like place where the wine is actually made. Our host was aged, at that time, ninety-two years old. He had inherited his vineyard in his twenties and had continued to use the wooden presses and other tools which had come to him with the vineyard. He complained a little that it was now sometimes difficult to find a man willing to work with traditional tools which were, he conceded, often heavier and not so easy to handle as newly-designed equipment. He was convinced the wooden implements made the better wine and he did not repine too much about the difficulty of finding help as he only needed one man. There is in Beaune, not far away, a fascinating museum on the making of wine and we had rather the impression of being in a part of it, actually given over to work, as we looked over the presses with the working parts worn black and shiny with use. Above was the *vigneron*'s house and, as the land sloped steeply upward on that side, the lower rooms of the house gave on to an ancient orchard with gnarled trees burdened to the ground with russet apples. Here in a flagged living-room next door to the kitchen, with daubs representing his rubicond early nineteenth-century ancestors hung round the walls, our host sat himself down gratefully and rather heavily and poured us out glasses of smooth old *ratafia* of his own making. He had some handsome pieces of furniture of solid country design and mostly made of walnut which had been lovingly polished every week since they had been bought. He was proud of them and he patted them affectionately as he introduced them to us.

'Now, this cupboard,' he would say, 'was made for my great-grandmother and was part of the dowry she brought with her when she married.'

As he passed one piece he opened a drawer and took out a silver *tastevin* for each of us and, poking into the pocket of the shapeless old grey cardigan which hung down to his knees, to make sure that he had got his own *tastevin* on him, he led the way across to the cellars. He saw me reading the inscription on my *tastevin* and twinkled at me. 'That was won by an uncle of mine,' he said. 'We never buy *tastevins* in my family. We win them at shooting matches.'

Down in the cellar he pottered about looking for a *pipette* and, removing in turn the little hatches which closed the barrels, sucked some wine up into it and squirted a little into each of our *tastevins*, on each occasion announcing the year. And then followed a long and rambling discussion. How one was plentiful but mediocre, another slow to mature but eventually would be good, another a marvellous year but alas, alas, so little in quantity, the whole interspersed with recollections extending over seventy years of great years, fine vintages, disastrous storms and the damage consequent on enforced absences mostly due to military service, illness or war. Through it all shone a love of his land, a pride in his skill and a conviction so deep as never to need expression, that the best trade on earth was to make good wine out of the Lord's good grapes in the most beautiful countryside He ever created. He was a very sweet old gentleman.

I have not yet found similar small *vignerons* in Italy partly because I do not yet have the right introductions or the time to follow up those I have. The best wines are of course always saved for the favourite customers who are known to appreciate them. Contacts have to be nursed and knowledge of methods and techniques acquired. One has to win one's way into the confidence of these men and it all takes time. There are very good wines in the Trento and around Verona and even if nothing has come of a day's journey in search of them, the day itself is often a success because wine so often grows in enchanting places.

CHAPTER XII

THE GARDEN

I STARTED planning the garden in the expectation that the soil
would have sufficiently recovered by the spring of 1967 to permit
at least some planting.

The garden was pleasantly complicated to plan. It had to combine the English and South African gardens of my childhood
with Mediterranean, tropical and American overtones. It had to
be simple and easy to run and to provide a wide variety of fruit
and vegetables for the house.

The main design imposed itself. Starting from the house there is a walk along the lagoon with the wild tamarisk left straggling over the water but kept lopped so that one can see the fishing boats go by. A long narrow rose-garden separates the walk from Jeanne's orchard, so-called from the friend who gave me the trees. This orchard is designed to supply the house with fruit and to be, in time, a pleasant place to stroll in. It occupies a large square of ground running back as far as the vineyard. By the side terrace and convenient to the kitchen is the site of the herb garden.

At the back of the house, stretching from the terrace towards the vineyard, there is a lawn and behind it a little copse of lilac and cob-nuts, where height would interrupt the view of the Torcello *campanile* and, further along, mostly of birch and Italian poplars with two or three lime trees. Eventually this little wood will be populated by spring flowers and a variety of shrubs. A wide flower bed runs along part of the wood. The broad path which eventually leads to the landward entrance of the vineyard and which separates it nearer the house from the orchard is lined with almond trees matched by apricots and cherries on the other side.

I was very much aware that I knew nothing about gardening, though I had been an eager if somewhat mercenary gardener as a child. Looking back on that time I think that my garden awoke my commercial instincts as much as it satisfied my liking for growing things. When I was about twelve we had a large and rambling garden and part of it was given over to me at my request. As my finances, then as now, were always in a deplorable state, I hit upon the notion of growing lettuces for my family. These I sold to my mother at a ha'penny a time and, as we had salad for lunch every day whenever they could be procured, this became quite a useful source of revenue. My garden eventually produced lettuces quite as good as the beautiful round, full-hearted ones with a little tinge of pink along the outer leaves which the Indian vegetable sellers used to bring to the house. Two or three times a week a slender young Indian woman would appear with a wide shallow basket balanced on her head which she lowered on to the kitchen steps for my mother's inspection. She had great taste in composing the still life in her basket where tomatoes, lettuces, peas (so oddly sold by the pint measure), sleek fat white leeks, aubergines and peppers contributed their different colours and textures. I now realize that I must have grossly under⁄ cut her prices. It is fair to say that I did not understand this at the time. In any case my need was very great. My weekly income amounted to threepence and I had the greatest difficulty in making ends meet. In the first place I needed to buy books and this was before the days of the sixpenny Penguins. Then I had a bicycle to keep up and in those days the diet in decent professional families was distinctly deficient in such necessary nutrients as fudge, liquorice and jelly babies. In these circumstances I was naturally forced into commercial practices. I found a willing, indeed complaisant, collaborator in my grandfather who, when I had got up specially early to pick, tie up in bunches and arrange in baskets the blossom which grew in such profusion on the flowering bushes in the garden, took them on my behalf to the weekly market in the town while I went to school. They were good for threepence a bunch and he never levied a commis⁄ sion. Another convenient source of income was my family's liking for cake. I had been taught how to make cakes at school and every Saturday afternoon I made a large batch, my mother of course providing the ingredients. The going price was one shilling and sixpence for making two victoria sandwiches with fillings, two large rounds of shortbread and what my grandmother would have called a sufficiency of scones. And so by hard work and letting no opportunity slip, I made ends meet, but what I really enjoyed was the gardening.

As I planned my garden in Torcello I came to recognize that a garden is one of the most usually transported items of one's national baggage. William III brought to his palace at Kensington a Dutch sunken garden to the lasting pleasure of Londoners. To him it must have been a whiff of home in a foreign land. Perhaps gardens are as indispensable to the émigré as his dearest prejudices. Perhaps, more innocently, one takes one's garden with one because it is so inextricably mixed with recollections of one's childhood.

Certainly, I have taken great care to plant a table grape called, in Italy, *regina*, which is the nearest I have been able to find to the oblong, sweet, fleshy golden grape known in the Cape Province as *harnepoort*. This was a conscious search on my part but I do not think that making a garden of a recognizably national kind is usually a conscious effort. It is more likely that when one makes a garden one tends to reproduce the gardens of one's childhood. I remember one spring walking along the road that runs round the base of the old town walls of Vézelay in Burgundy. Below me there was a small house with a wide lawn running up to the windows with daffodils growing under the old apple trees. The hedge enclosing it was neatly trimmed and the beds in the kitchen garden were dug all ready for planting. Within view of the windows there was a bird bath and beside it a ledge with crumbs for the birds. I looked at it in great puzzlement. It was at once very familiar, and quite wrong, and when I compared it with its neighbours the reason suddenly became clear to me. It was an English garden. I never discovered who lived there but it was inconceivable that he should not be English. In Italy, at the turn of the century, Captain Neil MacEacharn planted a superb garden at his house which he called the Villa Taranto in honour of his ancestor Lord Nelson, at the northern end of Lake Maggiore. He collected plants from all over the world and with great skill laid out in the hilly, sheltered land about the villa a recognizably English garden in spite of the exotic character of some of the trees and shrubs. He left his villa and garden to the Italian State as a horticultural research station and training centre. No visit to Lake Maggiore is complete without the Villa Taranto. In the first place it makes a remarkable contrast with those other, quite different but also magnificent, gardens in the Borromeo islands lower down the lake. Then one is shown round by a set of very knowledgeable gardeners who answer one's questions and provide one with specific and accurate information.

If the garden one makes is a reflection of one's national background, the notion of it is much influenced by one's reading as a child and the two become inextricably entwined. We gather what

a garden should be from Shakespeare and Herbert and Marvell.
One allows Leigh Hunt in less distinguished verse to prompt one
about gardens:

> Saying all one feels and thinks
> In clever daffodils and pinks;
> In puns of tulips, and in phrases,
> Charming for their truth, in daisies.

Even less wisely one allows O'Shaughnessy to remind one:

> I, too, will something make
> And joy in the making
> Though on the morrow it seem
> But the empty words of a dream
> Remembered on waking.

Surely in the vast savannahs of English poetry this must have been
said—if it had to be said at all—less rebarbatively. But it has
become part of one's baggage because one learnt it as a child.

The links between gardens and the people who furnish a child's
imagination are also tenacious. Washington chopping down his
cherry tree is more vivid than Washington the father of his country. Wordsworth, for me, sits for all time absurdly surrounded by
daffodils as though picnicking in a market garden in the Scilly
Isles, although I know it was Dorothy who saw them in the hills
of the North. Montaigne paces beneath his stately chestnuts,
George Sand takes a midnight dip in the pond of the rather dank
garden at Nohant (and Chopin, of course, catches a chill). Over
a bristly hedge one sees Colette driving her donkey briskly along
a Burgundian lane. Newton ponders on the apple which falls on
his head in the garden at Queens'. The young Milton, with lanky
fair hair drooping over his shoulders just like any junior member of
the college today, thinks out Comus under the mulberry in
Christ's College while Lawrence Sterne chortles over the more
indecent classical authors under the walnuts not far away in Jesus'.
So, with all this baggage, what could I make except an English
garden?

Making a garden takes time. The trees have to grow and so do one's notions of what one wants to do. One is planning for a distant future and becomes side-tracked on the way. The tenses in this chapter keep getting confused and this is a fair indication of the confused state of mind that making the garden induced. One has to think of the present and the future at the same time: the oak and the lettuce are both elements in the time scale of a garden. Making a garden needs foresight: it seems largely a question of leaving spaces in the right places, which is a very difficult thing to do. It needs patience: Hidcote, Sissinghurst, the Villa Taranto were all begun in this century and, though my garden will never reach these heights, they are tremendously encouraging. It needs lucky encounters: a new rhododendron from a friend's garden in Southborough, a magnolia from Oxford, a dogwood from Princeton, a strawberry tree from the gardens on the Borromeo Islands in Lake Maggiore, and the far-off recollection of a hillside in the Cape Peninsula.

I set off scientifically enough by having a soil analysis made and this led, as the best intentions so often do in gardening, to immediate disappointment. It became clear that it was out of the question to try to grow rhododendrons—except in a tub and who wants to grow a rhododendron in a tub? I thought of the rhododendrons at the Villa Taranto and wondered whether this was why Captain MacEacharn had sold his *palazzo* in Venice and gone to live in the hills. I took one look at the lagoon and sadly decided to forgo rhododendrons.

I continued to dazzle myself with science by finding out what grass had been used for the lawns at the Villa Taranto. I knew that Captain MacEacharn had discovered a grass in New Zealand which at the Villa gives a good shorn surface which, with some watering, survives the Italian sun. It is *eragrostis capillaris*. With the help of the Penguin book on lawns and against a background of jeers from my Italian friends who said it was madness to attempt to make a lawn which would surely be shrivelled by heat in the summer and by frost in the winter, Giulindo and I made a very fine lawn. Now four years later it is no longer a fine lawn, partly because I did not know enough about either the grass or about lawn-mowers. I have now learnt that *eragrostis capillaris* must be kept very short if it is to be prevented from making a mat of old, dead leaves on top of which sits the living part of the plant. A roto-mower does not keep it short enough. The other difficulty arose from my being away so much. Giulindo had never seen a lawn and my instructions about close cutting seemed to him an aberration. His notion of a lawn seemed to me more like a field of young winter wheat sprouting about two inches high. What with my supplying the wrong sort of mower and Giulindo the wrong idea of a lawn, *eragrostis capillaris* has made itself a fine eiderdown to sit on. And now, naturally, it is going bald in patches. When Torcello becomes my permanent base we shall have to make a new lawn. In the meantime *eragrostis capillaris* has strangled all the weeds which grew there in such profusion. The failure with the lawn I have put down to experience. In gardening, perhaps more than in other activities, it seems one must make one's own mistakes.

In the meantime I had planted Jeanne's orchard with a wide range of trees designed to stock the house with fruit, fresh, frozen, preserved or jammed throughout the year. There are cherries (early and late), apricots (early and late), peaches (both white and yellow, early and late), nectarines, plums (particularly greengages), six different kinds of fig, quinces, a mulberry, a medlar and a kaki because I had never seen one; nuts in profusion (walnuts, almonds and cobs); apples, particularly Cox's Orange (which do well in the north of Italy) and cooking apples (which nobody but the English seem to know about), which had to come from England. France supplied her great specialities, pears and greengages. So far there is only one problem. The medlar took to Torcello, grew by leaps and bounds and is already producing a vast number of fruit. But only I seem to like them and I can see myself becoming quite a bore pressing little pots of medlar cheese on unwilling friends. Soft fruit I am leaving until I go to live in Torcello as I fear a repetition of what happened with the lawn. Similarly all work in the garden has been postponed except for a number of shrubs which were planted in the autumn of 1967 as soon as the soil seemed to have recovered from the salt deposited by the flood. They have survived the remaining salinity in the soil.

For all these reasons the garden looks a little stark. It is still only a skeleton of a garden, populated (except for a few straggly old trees and the robust tamarisk growing along the retaining banks) by little trees and shrubs and, in the summer, by a few annuals broadcast here and there. At this stage I observe my garden with the eye of faith. I see not what I see but what I hope I will see in the future. The herb garden is an example. By the house, convenient to the kitchen, I see a herb garden, with some broken-down statuary amongst the plants, surrounded by a little wall built of old bricks from the ruined cottage. Pinks and a gilly flower or two cohabit with thyme and marjoram and mingle whiffs of England and Greece. It will all come with time.

While I wait I ponder the two remaining problems of the garden. One is labour and the other is me. Giulindo and his various relatives who come to help in the garden are all highly skilled. The vineyard is demonstration enough of that. The problem arises from their bashfulness in dealing with flowers and their ruthlessness in uprooting plants they do not recognize. Giulindo constantly tells me I must give him detailed orders in the garden because he has no knowledge. 'I am the specialist in the vineyard,' he says. 'I can prune vines and peaches and can judge exactly what care they need to produce good crops. And look at those beautiful artichokes! The *signorina* knows nothing about the vineyard. The *signorina*, however, is a specialist in the garden and I will do what she tells me. The result is her responsibility,' he adds, rather ominously. The difficulty, of course, is that he only does as much of what I tell him as he thinks sensible even when I am there. When I am not he must take responsibility and I am glad that he does so, even if calamities occur. He thinks growing flowers is a waste of time. They do not bring in any money. He has always refused to take flowers from the garden to his wife, though she is very glad to have them, and I have never been quite sure whether this is because he resents the amount of land given over unproductively to the garden or whether carrying flowers is in some way a reflection on his virility. Most of the work in the garden is done by women. They are quite pleased to grow flowers and inspect new ones with interest. It is their intention to obey their instructions implicitly but once they have a hoe in their hand they lay about them with expedition and skill and uproot every, thing except the few flowers, such as cosmos, larkspur, snap, dragons and sweet william (the whole island now seems to have adopted *dolce guilielmo* as the name of what is properly called *garofani cinesi* or sometimes *bei uomini* in Italian—so are languages extended) which they grow in their own gardens. Once I returned from Venice to be informed that, as they had finished their appointed jobs, they had 'put a little order' into a flower bed. I was proudly shown a perfectly empty, neatly raked section. Clearly the work had been carefully and lovingly done. Delphiniums, aubretia, perennial candy tuft and some very precious thrift be, longing to my mother, had all had their heads cut off and been cast into the lagoon. It was at this point that I decided that it was a useless expenditure of nervous energy and time and money to think of stocking the garden with perennials until I myself could give much more time to it.

In all honesty I have to concede that my deficiencies arose from exactly the same cause as theirs. In both cases it was a too re, stricted range of knowledge. They threw out plants they did not know. I refused to plant many which are handsome and would thrive in Torcello because I did not know them and was con, sequently prejudiced against them. I have now become a great reader of seedsmen's catalogues and visitor to their nurseries and after this enforced respite I hope the garden will advance more rapidly and with a greater variety of plants suited to their en, vironment.

So I sit and look at the garden as it will be and watch the bees gathering honey from the dog-roses. They do not deserve their reputation for orderly work. They visit the flowers unsystematic, ally, revisiting some constantly and as constantly missing out their neighbours, revisiting, missing out, hesitating, flying agitatedly back, forwards, sideways, up, down, pouncing, revisiting, missing out the best again. When they do settle they energetically rummage and ferret about and thrust in their heads and use wings and feet without a moment's repose. Presumably they rush from bush to bush as they rush from flower to flower. The energy they use to produce a pound of honey should be enough to drive a train. How exhausting! Surely they need a time and motion study. And how limiting! What a frightful life eternally pursuing honey within the narrow confines of a bush! I find myself reflecting that I am unlikely to become a bee, although not through any virtue of my own.

The fact of having lived all my life emotionally astride frontiers and of having been frequently uprooted for my job—or for my pleasure—has created a kind of built-in restlessness. Even now if I stay much longer than six weeks in any one place I begin to feel the need to go somewhere else, even if only for a weekend. This is why I enjoy so much my present triangular life based on Paris, London and Venice, two great capitals and one of the crossroads of Europe and each having its own flavour. Where, except Paris, would you find a shop given over entirely to snails? Where, except London, is the theatre really satisfactory? Where, except Venice, is one living in every century of European civilization at once, and sometimes with luck even in the twentieth century?

This restlessness is not so much a personal idiosyncrasy as a phenomenon of our times. More and more of us live a semi-nomadic existence, moving our homes from time to time or making regular sorties abroad from a permanent base. Many of my colleagues who have spent their whole working lives moving from place to place tell me that when they retire they will buy a house in a quiet spot and never move again. I wonder if they will be able to sit still in one place for twelve months at a time. I wonder, looking back on the retired Indian Army officers and colonial administrators who used to enthrall me with their stories, whether their constant harping on their past (because, to anything but the *tabula rasa* of a child's mind, it could be nothing else), whether this was not a way of sublimating an ache to go away. For myself I know that I could never live fixed in one place, not even a place like Venice, where almost everyone that one knows comes to or through at some point in their lives. I also know that I no longer wish to live engulfed in one national society, neither my own nor one that I am as fond of as that of France or Italy.

Although I had for years thought that one day I would live in the lagoon I had to take the actual decision to buy a house there in rather a hurry. I still ask myself rather anxiously whether this is really where I wanted my base to be. I do not know. What I do know is that I always arrive at the Ca' S. Tomà with a sense of homecoming and that I am equally pleased to go away again. In Auden's words it is 'a place I may go both in and out of'.

In spite of my comings and goings I have a great affection for this place, and as it has no roots in the past I welcome mooring lines to the future. There is the garden and the vineyard which will grow lovely with time and there is a hint of interest in the over-heard comment of a little boy: 'I'm glad she will give it to us when she is dead.'

INDEX

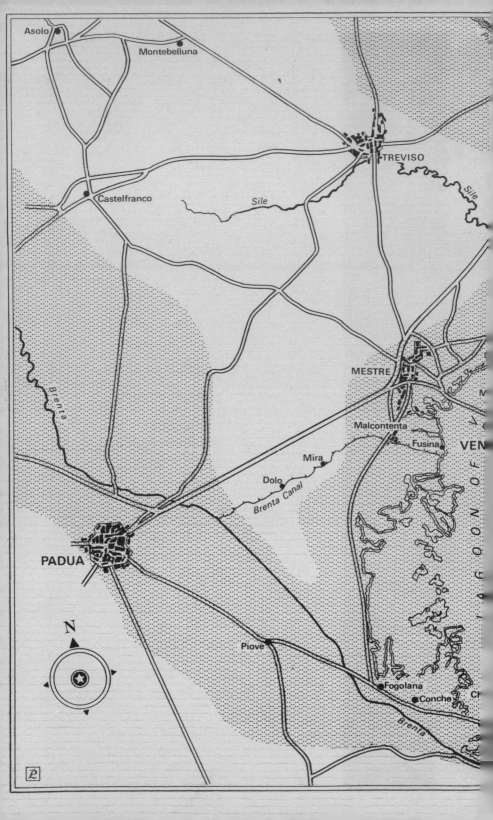